Hiram Mattison

A High-School Ascronomy

In Which the Descriptive, Physical and Practical Are Combined

Hiram Mattison

A High-School Ascronomy
In Which the Descriptive, Physical and Practical Are Combined

ISBN/EAN: 9783337812881

Printed in Europe, USA, Canada, Australia, Japan

Cover: Foto ©berggeist007 / pixelio.de

More available books at **www.hansebooks.com**

A

HIGH-SCHOOL ASTRONOMY:

IN WHICH THE

DESCRIPTIVE, PHYSICAL, AND PRACTICAL

ARE COMBINED,

WITH SPECIAL REFERENCE TO THE WANTS OF

ACADEMIES AND SEMINARIES OF LEARNING.

BY HIRAM MATTISON, A. M.,

LATE PROFESSOR OF NATURAL PHILOSOPHY AND ASTRONOMY IN THE FALLEY SEMINARY; AUTHOR OF THE PRIMARY ASTRONOMY; ASTRONOMICAL MAPS; EDITOR OF BURRITT'S GEOGRAPHY OF THE HEAVENS, ETC., ETC.

NEW YORK:
PUBLISHED BY MASON BROTHERS.
BOSTON: MASON & HAMLIN. PHILADELPHIA: J. B. LIPPINCOTT & CO.
CINCINNATI: W. B. SMITH & CO.
1863.

PREFACE.

The design of this work is to furnish a suitable text-book of Astronomy for academies and seminaries of learning. Though substantially a revised edition of the "*Elementary Astronomy,*" so extensive and important have been the additions and improvements, as to justify the adoption of a new title, and to warrant the hope that it will not only be found eminently suited to its purpose, but that it may now go on for years without further revision or alteration.

For juvenile learners, the "*Primary Astronomy*" may be preferred; and for advanced classes, who wish to study the constellations in connection with Mythology, the "*Geography of the Heavens*" should be chosen in preference to all others; but for all ordinary students, this intermediate work will be found sufficiently elementary on the one hand, and sufficiently extended on the other.

The work is now divided in three parts. After an Introduction, which consists of Preliminary Observations and Definitions, and occupies twenty pages, Part First is devoted to the Solar System—the sun, planets, comets, eclipses, tides, &c.; Part Second relates to the Sidereal Heavens—the fixed stars, constellations, clusters, and nebulæ; and Part Third to Practical Astronomy—the structure and use of instruments, refraction, parallax, &c. This department, so seldom introduced into text-books for schools, will be found especially interesting and valuable.

Besides embracing all the late discoveries in astronomy, under a strictly philosophical classification, the work is now thoroughly illustrated, by the introduction of diagrams into the pages, in connection with the text; and the adaptation throughout to the use of the blackboard, during recitation, cannot fail to be appreciated by every practical teacher.

The variety of type affords an agreeable relief to the eye of the student, and at the same time distinguishes the main text (which ought, in all cases, to be thoroughly understood before it is passed) from the less important matter, the more careful study of which may be left for a review. The suggestive topical questions at the bottom of the page complete the design.

On the whole, the work is believed to be a decided improvement upon the works heretofore in use in this department of study; and as such it is offered to the professional teachers of the country.

New York, Jan. 1, 1853.

ASTRONOMICAL WORKS

In the Author's Library, and more or less consulted in the compilation of the following pages:

A Cycle of Celestial Objects, for the use of Naval, Military, and Private Astronomers, &c. By CAPT. WM. HENRY SMYTH, &c. 2 vols. 8vo. London, 1844.

An Introduction to Astronomy, in a Series of Letters from a Preceptor to his Pupil, &c. By JOHN BONNYCASTLE, Professor of Mathematics, &c. 1 vol. 8vo. London, 1822.

An Introduction to the True Astronomy; or, Astronomical Lectures read in the Astronomical School of the University of Oxford. By JOHN KEILL, M. D., F. R. S., &c. 1 vol. 8vo. Dublin, 1793.

Astronomy Explained, upon Sir Isaac Newton's principles, &c., &c. By JAMES FERGUSON, F. R. S. 1 vol. 4to. London, 1764.

The Elements of Physical and Geometrical Astronomy. By DAVID GREGORY, M. D., late Sullivan Professor of Astronomy at Oxford, &c. 2 vols. 8vo. London, 1726.

Astronomy, in Five Books. By ROGER LONG, D. D., F. R. S., &c., University of Cambridge. 2 vols. 4to. Cambridge (Eng.), 1742.

Astronomia Carolina, &c., by THOMAS STREET; and *A Series of Observations on the Planets*, chiefly the Moon, &c., by DR. EDMUND HALLEY. 1 vol. 4to. London, 1716.

Astronomical Lectures, read in the Public School at Cambridge (Eng). By WILLIAM WHISTON, M. A., Professor of Mathematics, &c. 1 vol. 8vo. London, 1728.

The Wonders of the Heavens; a popular view of Astronomy, &c. By DUNCAN BRADFORD. 1 vol. royal 4to. New York, 1843.

Popular Lectures on Science and Art, &c. By DIONYSIUS LARDNER, F. R. S., &c. 2 vols. 8vo. New York, 1846.

Outlines of Astronomy. By SIR JOHN F. W. HERSCHEL, Bart., K. H., &c. 1 vol. 8vo. Philadelphia, 1849.

Phenomena and Order of the Solar System, and Views of the Architecture of the Heavens. By J. P. NICHOL, F. R. S. E., &c. 2 vols. 12mo. New York, 1842.

The Practical Astronomer, &c. By THOMAS DICK, LL.D. 1 vol. 12mo. New York, 1846. Also, "Celestial Scenery," and "The Sidereal Heavens," by the same author.

The Planetary and Stellar Worlds. By PROF. O. M. MITCHEL. 1 vol. 12mo. New York, 1849.

An Elementary Treatise on Astronomy, &c. By WILLIAM A. NORTON, A. M. 1 vol. 8vo. New York, 1845.

An Introduction to Astronomy, &c. By DENISON OLMSTED, A. M. 1 vol. 8vo. New York, 1844. Also, *Letters on Astronomy*, and *Life and Writings of Ebenezer Porter Mason*, by the same author. 2 vols. 12mo.

The Solar System; or, the Sun, Moon, and Stars. By J. R. HIND, Director of Mr. Bishop's Observatory, Regent's Park, London. 1 vol. 12mo. London, 1852.

A Pictorial Display of the Astronomical Phenomena of the Universe, &c. By C. F BLOUNT. 4to. New York, 1844.

The Recent Progress of Astronomy, &c. By ELIAS LOOMIS, Professor of Mathematics, &c. 1 vol. 12mo. New York, 1850.

Annual of Scientific Discovery, &c. By DAVID A. WELLS, A. M. 1 vol. 12mo. Boston, 1852.

The Sidereal Messenger; a Monthly Journal, devoted to Astronomical Science. By O. M. MITCHEL, A. M. (Now discontinued.)

Also, Astronomical Lectures, by ARAGO, LARDNER, MITCHEL, and NICHOL; and Elementary Treatises by BURRITT, KENDAL, BARTLET, MCINTIRE, ABBOTT, OSTRANDER, BLAKE, HASLER, SMITH, CLARK, VOSE, TYLER, COMSTOCK, HASKINS, RYAN, WILKINS, KEATH, &c., &c.

CONTENTS.

INTRODUCTION.

PRELIMINARY OBSERVATIONS AND DEFINITIONS.

	PAGE
CHAP. I.—ORIGIN AND HISTORY OF THE SCIENCE.	
Ptolemaic Theory of the Structure of the Universe	12
The Copernican System	13
II.—DEFINITIONS.	
Solids, Surfaces, &c.	16
Spheres, Hemispheres, and Spheroids	17
Lines and Angles	19
Of Triangles	20
Circles and Ellipses	21
The Terrestrial Sphere	23
The Celestial Sphere	25
First Grand Divisions of the Universe	28

PART FIRST.

THE SOLAR SYSTEM.

CHAP. I.—THE PRIMARY PLANETS.	
Classification of the Solar Bodies	29
Names of the Primary Planets	31
Explanation of Mythological Signs	32
Distances of the Planets	36
Light and Heat of the Planets	38
Magnitude of the Planets	40
Density	41
Gravitation	42
Periodic Revolutions of the Planets	44

CONTENTS.

	PAGE
Chap. I.—Hourly Motion of the Planets in their Orbits	45
Centripetal and Centrifugal Forces	45
Laws of Planetary Motion	46
Aspects of the Planets	48
Sidereal and Synodic Revolution	49
The Ecliptic, Zodiac, Signs, &c.	50
Celestial Latitude and Longitude	53
Mean and True Places of a Planet	54
Direct and Retrograde Motions	55
Morning and Evening Stars	57
Deviation of the Orbits of the Planets from the Ecliptic	58
Philosophy of Transits	60

II.—Primary Planets Continued.

Inclination of the Axis of the Planets, and its Effects	65
Rotation of the Planets upon their axes	69
Time	70
Equation of Time	72
Time, as affected by Longitude	76
True Figure of the Planets	77
Precession of the Equinoxes	80

III.—Telescopic Views of the Planets.

Mercury—Phases, Mountains, &c.	83
Venus—Phases, Mountains, Atmosphere	84
Mars—"Continents and Seas," Color, Snow Banks	86
The Asteroids—Color, Hazy Appearance	87
Jupiter—Oblateness, Belts, Moons	88
Saturn—Oblateness, Rings, Belts, Phases, Moons	89
Uranus—a Telescopic World, Satellites	94
Neptune—purely Telescopic, Satellite	95
Herschel's Solar System in Miniature	95

IV.—Seasons of the Different Planets, &c.

Cause of the Seasons—Mercury, Earth	97
Venus, Mars, Jupiter	98
Saturn and Uranus	99
Discovery of the different Planets	100

V.—Secondary Planets—The Moon.

Character and Number of the Secondaries	102
The Moon's Distance, Shape, Position of Orbit, &c.	103

CONTENTS.

	PAGE
CHAP. V.—Magnitude, Density, Revolution Eastward	105
Form of Lunar Orbit	107
Cause of the Moon's Changes	109
Natural appearance—Same side always toward us	111
Moon's Librations—in Latitude and Longitude	112
Telescopic Appearance of the Moon—Lunar Mountains	113
Finding the Longitude by the Moon's place	115

VI.—ECLIPSES OF THE SUN AND MOON.

Philosophy of both	116
Law of Shadows	117
Why not two Eclipses every Lunar Month	118
Why Solar pass *eastward* over the Sun, and Lunar *westward* over the Moon	119
Ecliptic Limits—Umbra and Penumbra	120
Why all Central Eclipses not total	122

VII.—SATELLITES OF THE EXTERIOR PLANETS.

Satellites of Jupiter—Distances, Periods, &c.	124
Eclipses of Jupiter's Moons—Immersions and Emersions	126
Moons of Saturn—Why seldom eclipsed	127
Satellite of Neptune	129

VIII.—NATURE AND CAUSE OF TIDES.

Description of Tides, Causes, &c.	130
Spring and Neap Tides	134

IX.—OF COMETS.

Name, Parts, Orbits, &c.	136
Magnitudes, Velocity, Temperature, Periods	140
Numbers, Physical Natures, &c.	143

X.—THE SUN.

True Figure, Spots	146
Physical Constitution, Temperature	151
Zodiacal Light	153
Sun's Proper Motion in Space	155

XI.—MISCELLANEOUS REMARKS UPON THE SOLAR SYSTEM.

Nebular Theory of its Origin	156
Were the Asteroids originally one Planet?	159
Are the Planets inhabited by rational beings?	161

PART SECOND.
THE SIDEREAL HEAVENS.

CHAP. I.—THE FIXED STARS.

	PAGE
Classification of the Stars	166
Number of the Stars	168
Distances of the Stars	170

II.—DESCRIPTION OF THE CONSTELLATIONS.

Nature, Origin, Classification	172
Visible in October, November, and December	174
" January, February, and March	177
" April, May, and June	180
" July, August, and September	183

III.—DOUBLE, VARIABLE, AND TEMPORARY STARS, &c.

Stars Optically and Physically Double	187
Binary Systems	189
Variable or Periodical Stars	194
Temporary Stars—New and Lost	196

IV.—CLUSTERS OF STARS AND NEBULÆ.

Pleiades, Hyades, &c.	199
Nebulæ—Resolvable, Irresolvable, Annular, &c.	201
Planetary, Stellar, &c.	203
Star Dust, Milky Way	206

PART THIRD.
PRACTICAL ASTRONOMY.

CHAP. I.—PROPERTIES OF LIGHT.

Refraction of Light	211
Atmospherical Refraction	214
Refraction by Glass Lenses	216

II.—TELESCOPES.

Refracting Telescopes	221
Reflecting Telescopes	231
Transit Instrument	235
Mural Circle	236
Parallax	237
Meteors and Meteoric Stones	239

INTRODUCTION.

PRELIMINARY OBSERVATIONS AND DEFINITIONS.

CHAPTER I.

ORIGIN AND HISTORY OF THE SCIENCE.

1. SCIENCE is knowledge systematically arranged, so as to be conveniently taught, easily learned, and readily applied.

2. ASTRONOMY is the science of the heavenly bodies—the Sun, Moon, Planets, Comets, and Fixed Stars.

The term *astronomy* is from the Greek *astron*, a *star*, and *nomos*, a *law*; and signifies the laws or science of the stars.

3. Astronomy is divided into *Descriptive*, *Physical*, and *Practical*.

Descriptive Astronomy includes the mere *facts* of the science, irrespective of the *causes* of the phenomena observed, or of the *means* by which the facts were ascertained.

Physical Astronomy explains the *causes* of the various phenomena observed, as of Day and Night, the Seasons, Eclipses, Tides, &c.

Practical Astronomy relates to the *means* for acquiring astronomical knowledge by the use of instruments, and by mathematical calculations.

These three departments have arisen, one after the other, in the order in which they are here stated. At first a few facts and phenomena were observed, but the *causes* were unknown. Next some of the causes were investigated one by one; and, finally, *instruments* were invented for measuring distances, altitudes, &c.; data for *calculations* were obtained; and thus arose the department of Practical Astronomy.

---

1. Define the term *Science*.
2. What is *Astronomy*? (From what is the term derived?)
3. How is astronomy divided? Descriptive? Physical? Practical? (State the order in which these departments have arisen.)

4. Astronomy has long been regarded as the most sublime of the sciences, eminently calculated to illustrate the wisdom, power, and goodness of God; to elevate and expand the human mind, and to fill it with exalted views of the *Creator*—

"The glorious Architect who built the skies."

1. "The greatest men of all ages have pronounced this science to be the most sublime and surpassing of all that can be tested by human genius, and to be worthy of a life of study."—*Smyth's Celestial Cycle.*
2. "Our very faculties are enlarged with the grandeur of the ideas it conveys, our minds exalted above the low-contracted prejudices of the vulgar, and our understandings clearly convinced, and affected with the conviction of the existence, wisdom, power, goodness, and superintendency of the SUPREME BEING!"—*Ferguson.*
3. So remarkably does this science exhibit the glory and majesty of God, by its astounding revelations of His *works*, that it almost necessarily tends to fill the mind with awe and reverence. It was in view of this tendency that the poet Young said,

"An undevout astronomer is mad."

4. To the moral influence of the contemplation of the heavens, we have frequent reference in the sacred Scriptures. "The heavens declare the glory of God; and the firmament showeth his handy-work." (Psalm xix. 1.) "When I consider thy heavens, the work of thy fingers; the moon and stars, which thou hast ordained; what is man, that thou art mindful of him? and the son of man, that thou visitest him?" (Psalm viii. 3, 4.)

5. Astronomy is probably the *most ancient* of all the sciences. Some of the Chaldean observations date as far back as 2,250 years before Christ, or only 98 years after the Flood! Laplace speaks confidently of Chinese observations 1,100 B. C.; and Mr. Bailly, an English astronomer, fixes the time of a conjunction of Mars, Jupiter, Saturn, and Mercury, mentioned in Chinese records, at 2,449 years before Christ.

1. The ancient Chinese astronomers and mathematicians were held to a fearful responsibility for the correctness of their calculations. In the reign of the Emperor Chouang, his two chief astronomers, *Ho* and *Hi*, were condemned to death for neglecting to announce the precise time of a solar eclipse, which took place B. C. 2,169.
2. The Holy Scriptures, some parts of which are very ancient, contain several allusions to the science of astronomy. In the first chapter of Genesis we have an account of the *creation* of the Sun, Moon, and Stars. "And God said, Let there be lights in the firmament of the heaven, to divide the day from the night, and let them be for signs, and for seasons, and for days and years. And let them be for lights in the firmament of the heaven, to give light upon the earth: and it was so. And God made two great lights; the greater light to rule the day, and the lesser light to rule the night: he made the stars also." Verses 14–16.
3. In the book of Job, written 1,500 years before Christ, we read of several constellations that bear the same names now that they did three thousand years ago. "Which maketh Arcturus, Orion, and Pleiades, and the chambers of the south." (ix. 9.) Again. "Canst thou bind the sweet influences of Pleiades, or loose the bands of Orion? Canst thou bring forth Mazzaroth in his season? or canst thou guide Arcturus with his sons?" (Chap. xxxviii. 31, 32.)

---

4. How astronomy regarded? (Smyth? Ferguson? Young? Scriptures?)
5. What of antiquity of astronomy? Chaldean and Chinese observations? (Responsibility of Chinese astronomers? Ancient Scriptural allusions?)

6. The first astronomers were *shepherds* and *herdsmen*, who were led to this study by observing the movements of the sun, moon, and stars, while watching their flocks from year to year in the open fields.

ANCIENT ASTRONOMERS OBSERVING THE HEAVENS.

7. *Thales*, one of the seven wise men of Greece, was the first regular teacher of Astronomy, B. C. 600. The next was *Anaximander*, a disciple of Thales, who succeeded him as head of the school at Miletus, B. C. 548. He asserted the true figure of the earth, and seems to have had some idea of its daily revolution.

<small>Anaximander is supposed to have been the first who constructed globes and maps. He taught that the moon shines by reflection, and in several other respects advanced beyond the knowledge imparted by his distinguished tutor.</small>

8. *Pythagoras*, another Greek philosopher, who founded the school of Croton, B. C. 500, greatly enlarged the science. He first gave form to the vague ideas that the sun was in the center of the planetary orbits, that the earth floated unsupported in space, and that the distant stars were *worlds*, and probably inhabited.

<small>"It was Pythagoras," says Smyth, "who taught, in fact, the system which now immortalizes the name of Copernicus." But he adds that his teachings were but "the conjectures of a sagacious mind, not possessed of the evidence requisite to give stability to its opinions." Pythagoras is said to have perished from hunger, in his old age.</small>

6. Who were the first astronomers? How led to this study?
7. Who first regular teacher of this science? How early? Who next?— and when? What correct notions did he seem to entertain? (For what else distinguished?)
8. Who next after Anaximander? What advances did he make in this study? (What does Smyth say of his teachings? What said of his death?)

9. *Ptolemy*, an Egyptian philosopher, taught astronomy in the second century of the Christian era. He adopted the theory that the earth was located in the center of the universe, that it was perfectly at rest, and that the sun, moon, and stars actually revolved around it, from east to west, as they appear to do, every twenty-four hours. This system is called, after its author, the *Ptolemaic Theory*.

PTOLEMAIC THEORY OF THE STRUCTURE OF THE UNIVERSE.

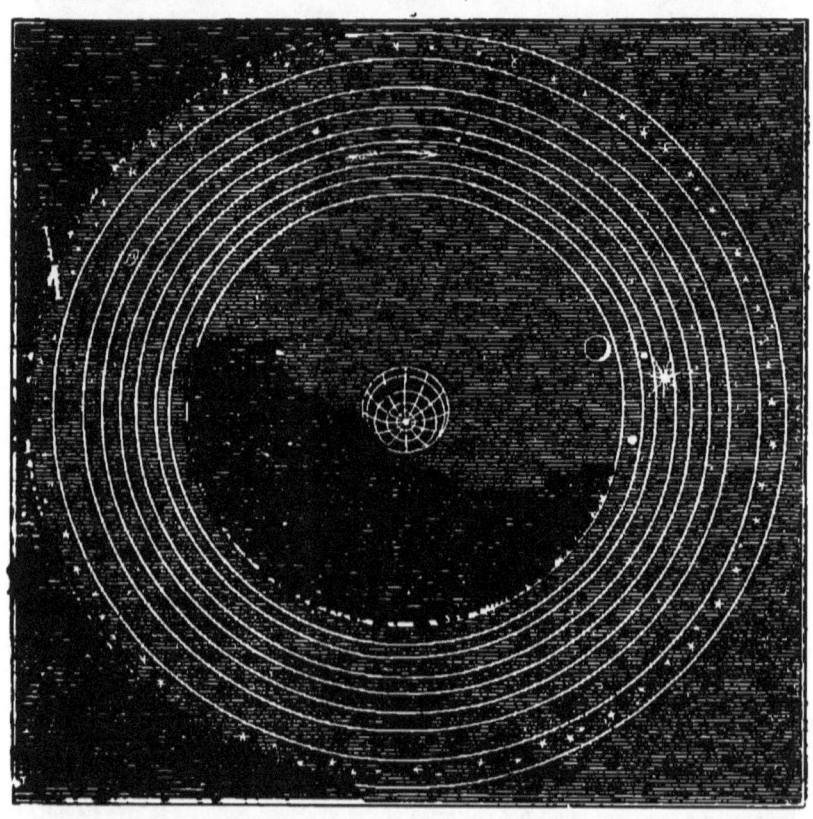

1. Ptolemy supposed the earth to be in the center of a system of crystalline arches, or hollow spheres, arranged one within the other, as represented in the cut. It is thought by some that he understood the spherical figure of the earth, and the cut is constructed upon this supposition. Ptolemy further supposed that the sun, moon, and stars were fixed in these crystalline spheres, at different distances from our globe; that the Moon was in the first, Mercury in the second, Venus in the third, the Sun in the fourth, Mars,

9. Who was Ptolemy?—and when did he flourish? Describe his theory. (How locate sun, moon, &c.? What absurdity did it involve, as it respects

Jupiter, and Saturn in the next three, and the fixed stars in the eighth. The ancients had no knowledge of Uranus or Neptune. This ponderous machinery was supposed to revolve from east to west around the earth, carrying with it the sun, moon, and stars, every twenty-four hours; and the spheres being crystal, the distant stars were visible through them.

2. If the sun was designed to enlighten and warm the different sides of our globe, the Ptolemaic method of effecting this object is most unreasonable. To carry the sun around the earth, to warm and enlighten its different sides, instead of having the earth turn first one side and then the other to the sun, by a revolution on its axis, would be like carrying a fire around a person who was cold, and wished to be warmed, instead of his turning himself to the fire as he pleased.

3. The Ptolemaic theory would require a motion inconceivably rapid in all the heavenly bodies. As the sun is ninety-five millions of miles from the earth, the entire diameter of his sphere would be one hundred and ninety millions of miles, and its circumference about six hundred millions. Divide this distance by twenty-four—the number of hours in a day—and it gives *twenty-five million miles an hour*, or sixty-nine thousand four hundred and forty-four miles per second, as the velocity of the sun! This theory would require a still more rapid motion in the fixed stars. It would require the nearest of these to move at the rate of nearly *fourteen thousand millions of miles per second*, or seventy thousand times as swift as light, in order to accomplish their daily course. But with all these difficulties in its way, the Ptolemaic theory was generally believed till about the middle of the sixteenth century, or three hundred years ago.

THE COPERNICAN SYSTEM.

10. About the year 1510, the ancient theory of *Pythagoras* was revived and improved by *Copernicus*, a Prussian astronomer, and has since been called, after him, the *Copernican System*.

1. The investigations of Copernicus were conducted between the years 1507 and 1530. In the latter year he finished his tables of the planets, and his great work, *The Revolution of the Celestial Orbs*; but he did not venture to *publish* his views till thirteen years after, or 1543, when he received a copy of it only a few hours before his death, and consequently never read it in print. It contains the old philosophy, interspersed with his own original and acute conceptions, and was received under very considerable opposition.—*Smyth*, vol. 1, p. 38.

2. Copernicus is generally regarded as the *discoverer* of the system which bears his name, but this is a popular error. There is abundant proof, notwithstanding the loss of his writings, that Pythagoras understood the leading features of what is now called the Copernican Theory.

11. The first prominent feature of the Copernican system is, that the earth is a *sphere* or *globe*, inhabited on all sides.

The evidence that the earth is a sphere or globe may be arranged and stated as follows:

1. Admitting that the sun, moon, and stars are *worlds*, the fact that they are *round*, as we see them to be, affords ground for the presumption, at least, that the earth also is round.

2. Water falling from the clouds is gathered into little globes or *drops;* and melted lead poured from the summit of a high tower assumes the form of globes, which, when cooled, are called *shot*. And the same law would cause a larger mass of fluid matter, if left undisturbed in space, to assume the same shape. But the Bible teaches that the

---

light and heat? What in respect to the motions of the heavenly bodies? Was such a theory ever generally believed? Till how recently?)

10. Who was Copernicus? For what distinguished? About what time? (What of his investigations? His work? Its publication? Character? What popular error noticed?)

11. State the first leading feature of the Copernican theory. (What proofs of its correctness? The first? Second? Third? Fourth? Fifth? Sixth? Seventh?)

whole earth was once in a fluid state—one vast drop—the substances now constituting the oceans and continents being indiscriminately mingled together. "And the earth was without form and void [*i. e.*, chaotic, confused, unorganized], and darkness dwelt upon the face of the *deep;* and the spirit of God moved upon the face of the *waters*. * * * And God said, Let the waters under the heavens be gathered together unto one place, and let the dry land appear: and it was so. And God called the dry land *earth*, and the gathering together of the waters called he *seas*."—Genesis i. 2, 9. 10. Up to this time there was no "earth," either as continents or islands, neither were there any "seas," but all the elements were mingled together; and a mass of fluid thus dropped into space, from the hand of the Creator, would be as certain to assume the form of a globe, as the melted lead from the shot-tower, or the water from the passing cloud.

3. The apparent elevation and depression of the North Star, as we approach toward or recede from it, shows that the surface of the earth is convex, or that the earth is a globe.

4. The fact that the tops of mountains are last seen as we recede from, or first as we approach, the sea-shore, proves that the surface of the water upon which we sail is convex; so when a ship is approaching the shore, the topmasts are always seen first, and the hull or body last. And when seamen wish to survey the horizon at sea to a great distance, in search of whale or other shipping, they "go to the mast-head," as they call it, from which point they can often discover objects that are entirely invisible from the deck of the ships.

5. If an aqueduct is to be constructed a mile long, so as to be filled with water to the brim at every point, it must be about eight inches *higher* in the middle than at the ends, so as to allow the surface of the water to conform to the convex figure of the globe. We say *higher*, not that it needs to be higher as determined by a water level, for a water level is convex, but higher as determined by a straight line drawn from one end of the aqueduct to the other. This definite knowledge of the curvature of water, even for small distances, shows that the earth's surface is convex—or, in other words, that the earth is spherical. (The curvature from a tangent line is 8 inches for one mile, from the point of contact; 32 inches for two miles; 72 inches for three miles, &c.)

6. When the moon falls into the shadow of the earth and is eclipsed, or, in other words, the earth gets into her sunlight, and throws its shadow upon her, the shadow is seen to be convex. We must either conclude, therefore, that the earth, which casts the shadow, is in the form of a dinner-plate, and is always kept sidewise, and the same side toward the sun (which we know is not the case); or that it is a globe, and casts a conical shadow, whatever its position.

7. The earth is known to be a globe, from the fact that ships are constantly sailing around it.

8. It is not certain whether Ptolemy admitted the earth to be a sphere or not. Some writers maintain that he rejected this doctrine, and others that he admitted it. In the "PRIMARY ASTRONOMY," page 8, the author has inserted a cut representing the Ptolemaic theory, with the earth *flat;* but in this work (page 12), where the same theory is represented, the earth is shown as a *globe*. In all other respects, the theory represented is the same in both works; and this is only a minor point in the system.

12. A *second* leading feature of the Copernican theory is, that the apparent revolution of the sun, moon, and stars westward every day, is caused by the revolution of the earth around its own axis, from west to east, every twenty-four hours.

That the heavenly bodies *appear* to revolve westward, is no proof that they are actually in motion. We often transfer our own motion, in imagination, to bodies that are at rest; especially when carried swiftly forward without any apparent cause, as when one travels in a steamboat or railway car; and when for a time he forgets his own motion. "Copernicus tells us that he was first led to think that the apparent motions of the heavenly bodies, in their diurnal revolution, were owing to the real motion of the earth in the opposite direction, from observing instances of the same kind among terrestrial objects; as when the shore seems to the mariner to recede as he rapidly sails from it, and as trees and other objects seem to glide by us, when, on riding swiftly past them, we lose the consciousness of our own motion." This remark would go to show that the revolution of the earth on its own axis was an original discovery with Copernicus.

12. State the *second* leading feature of the Copernican system. (Do not our own senses furnish proof that the heavenly bodies revolve westward daily? Why not? What remark from Copernicus? What does it seem to imply?)

13. A *third* feature of the Copernican theory is, that the sun is the grand center around which the earth and all the other planets revolve.

THE COPERNICAN SYSTEM.

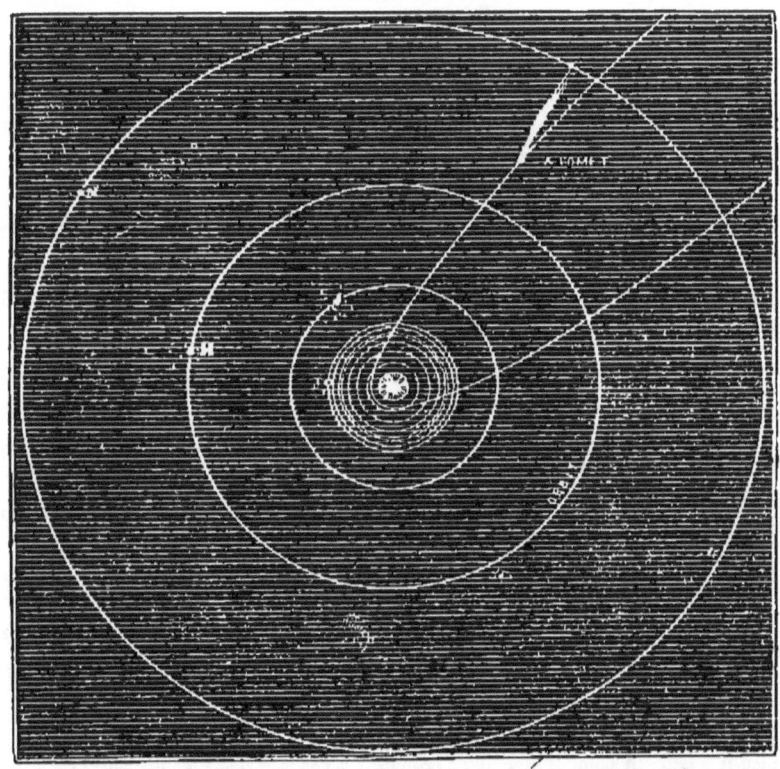

1. The above cut is a representation of the *Copernican Theory* of the Solar System. In the center is seen the sun, in a state of rest. Around him, at unequal distances, are the planets and fixed stars—the former revolving about him from west to east, or in the direction of the arrows. The white circles represent the *orbits*, or paths, in which the planets move around the sun. On the right is seen a *comet* plunging down into the system around the sun, and then departing. This is the *Copernican Theory* of the Solar System.

"O how unlike the complex works of man,
Heaven's easy, artless, unencumber'd plan!"

2. The truth of the Copernican theory is established by the most conclusive and satisfactory evidence. Eclipses of the sun and moon are calculated upon this theory, and astronomers are able to predict thereby their commencement, duration, &c., to a minute, even hundreds of years before they occur. We shall therefore assume the truth of this system without further proof, as we proceed hereafter to the study of the heavenly bodies.

---

13. State the *third* prominent feature of the theory of Copernicus. (Describe the cut. What additional evidence of the truth of this theory, as a whole?)

CHAPTER II.

DEFINITIONS.*

14. Solids, Surfaces, &c.

A *Solid*, or *Body*, is a figure having length, breadth, and thickness.

A *Surface* is the *outside* or *exterior* of a body, and has length and breadth only.

Surfaces are of three kinds—*Plane, Concave,* and *Convex.*

<small>A surface may also be *rough* or *smooth, hard* or *soft;* the above definition having reference only to the *general figure* of bodies.</small>

A *Plane Surface* is one that is perfectly flat or even, like the floor of a building, or the sides of a room.

<small>1. We may imagine what is called a *plane*, to extend off beyond the *plane surface* as far as we please; or, in other words, to be *indefinitely extended*. When a plane or a line is extended in this way, it is said to be *produced*.
2. An imaginary plane may exist where there is no *body* having a plane surface; or between two lines, like the place of a circle. A sheet of tin, laid across a small wire hoop, would represent the plane of that circle, in whatever position it might be held, whether horizontally, perpendicularly, or otherwise; and the place which the tin would pass through, if extended to the starry heavens, is the plane of that circle.
3. All objects which the tin would touch or *cut*, if extended outward to the heavens, or to infinity, are *in the plane* of the sheet, or the circle upon which it is laid. A point is in a plane produced, when the plane continued or extended would pass through that point.</small>

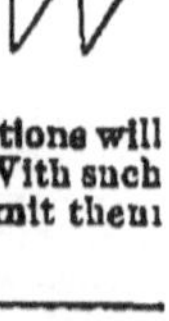

Parallel Planes are such as would never meet or cut each other, however far they might be extended.

<small>The two sides of a board, or two sheets of tin placed equidistant from each other at every point, represent parallel planes.</small>

<small>* To some who will use this work, many of the following diagrams and definitions will be superfluous, the substance of them being already sufficiently understood. With such students the judicious teacher will pass rapidly over the next ten pages, or omit them altogether.</small>

<small>14. Define a *solid*, or *body*—a surface. How many kinds of surfaces? (Any other distinctions?) What is a plane surface? (May a plane extend beyond the plane surface? May a plane exist where there is no body? Illustrate. What is a plane *produced?*) What are *parallel* planes? Perpen</small>

PERPENDICULAR PLANES.

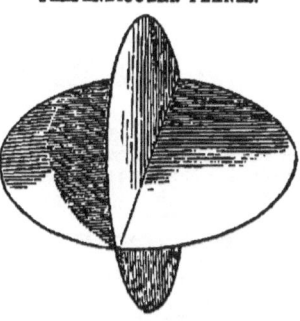

Perpendicular Planes are such as stand exactly upright upon each other, or cross each other at right angles.

<small>In the figure, one plane is placed horizontally, and the other perpendicular to it. They are therefore perpendicular *to each other*, however they may stand in relation to the observer.</small>

Inclined Planes are such as are inclined toward, and cut each other obliquely.

INCLINED PLANES.

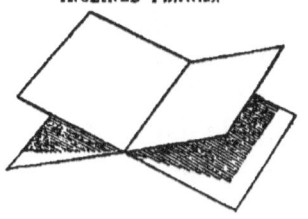

The *Angle of Inclination* is the angle contained between the two surfaces of the planes nearest each other.

<small>The spaces A and B in the adjoining cut represent the Angle of Inclination.</small>

The *Area* of a plane figure is the amount of surface contained therein.

A *Convex Surface* is one that is swollen out like the outside of a bowl.

CONVEX AND CONCAVE SURFACES.

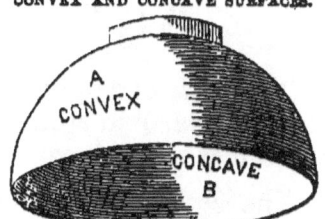

A *Concave Surface* is one that is hollowed out like the inside of a bowl.

15. SPHERES, HEMISPHERES, and SPHEROIDS.

A SPHERE.

A *Sphere* is a *globe* or *ball*, every part of the surface of which is equidistant from a point within, called its center.

<small>This is the ordinary definition; but in Astronomy, the term is applied to the apparent concave of the heavens, as if it were the actual concave surface of a hollow sphere.</small>

---

dicular? Inclined? What is meant by the angle of inclination? The *area* of a plane surface? Describe a *convex* surface—a *concave*.
 15. Describe a sphere—hemisphere—spheroid. (Derivation of spheroid?)

A *Hemisphere* is the *half* of a sphere or globe, or of the apparent concave of the heavens.

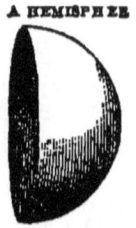
A HEMISPHERE

In Geography we often read of the Eastern and Western, and Northern and Southern hemispheres, but in Astronomy the term is only applied to the Northern and Southern portions of the heavens.

A *Spheroid* is a body *resembling* a sphere, but yet not perfectly round or spherical.

The term *spheroid* is from the Greek *sphaira*, a *sphere*, and *eidos*, *form*, and signifies sphere-like.

Spheroids are of two kinds—*Oblate*, and *Oblong* or *Prolate*.

AN OBLATE SPHEROID.

An *Oblate Spheroid* is a globe slightly flattened, as if pressed on opposite sides.

This is a difficult figure to represent upon paper. Should the pupil fail to obtain a correct idea, the Teacher will be at no loss for an illustration.

A *Prolate* or *Oblong Spheroid* is an elongated sphere.

This figure, like an Oblate Spheroid, admits of various degrees of departure from the spherical form. It may be much or but slightly elongated, and the ends may be alike or otherwise. A common egg is an Oblong Spheroid.

The *Axis* of a sphere is the line, real or imaginary, around which it revolves.

AXIS OF A SPHERE.

The *Poles* of a sphere are the extremities of its axis, or the points where the axis cuts the two opposite surfaces.

The *Equator* of a sphere is an imaginary circle upon its surface, midway between its poles, the plane of which cuts the axis perpendicularly, and divides the sphere into two equal parts or hemispheres.

Kinds of spheroids? Describe each. What is the axis of a sphere? What the *poles?* The *equator?* By what other name called? What a Less Circle? Meridians?

The equator of a sphere is sometimes called a *Great Circle*, because no larger circle can be drawn upon its surface.

GREAT AND LESS CIRCLES.

A *Less Circle* is one that divides a sphere into two unequal parts.

<small>In the cut, the circles are represented in perspective. The Great Circle embraces the *middle* of the sphere, where its full diameter is included; while the Less Circle passes around it between the Equator and the Poles, and is consequently "less" than the Equator.</small>

MERIDIAN.

Meridians of a sphere are lines drawn from pole to pole upon its surface.

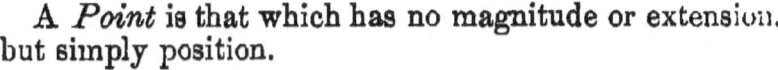

16. Lines and Angles.

A *Point* is that which has no magnitude or extension, but simply position.

<small>"The common notion of a point is derived from the extremity of some slender body, such as the extremity of a common sewing-needle. This being perceptible to the senses, is a *physical point*, and not a *mathematical point;* for, by the definition, a point has no magnitude."—Professor Perkins.</small>

A *Right Line* is the shortest distance between two points.

A RIGHT LINE.

A *Curve Line* is one that departs continually from a direct course.

CURVE LINE.

Parallel Lines are such as remain at the same distance from each other throughout their whole extent.

PARALLEL LINES.

Oblique Lines are such as are not parallel, but incline toward or approach each other.

OBLIQUE LINES.

When two lines intersect or cut each other, the space included between them is called an *Angle*.

AN ANGLE.

ANGLE.

<small>16. What is a point? (Physical? Mathematical?) A right line?—a curve line?—parallel lines?—an angle?—kinds of angles? Describe a right angle—an acute—an obtuse.</small>

Angles are of three kinds—namely, the *Right Angle*, the *Acute Angle*, and the *Obtuse Angle*.

Right Angles are formed when one right line intersects another *perpendicularly*, and the angles on each side are equal.

RIGHT ANGLES.

An *Acute* angle is one that is *less*, and an *Obtuse* angle one that is *greater*, than a right angle.

ACUTE AND OBTUSE ANGLES.

ACUTE.   OBTUSE.

17. OF TRIANGLES.

A *Triangle* is a plane figure, bounded by straight lines, and having only three sides.

Triangles are of six kinds—viz., *Right-angled, Obtuse-angled, Acute-angled, Equilateral, Isosceles,* and *Scalene.*

A *Right-angled Triangle* is one having one right angle.

The parts of a Right-angled Triangle are the Base, the Perpendicular, and the Hypothenuse.

RIGHT-ANGLED TRIANGLE.

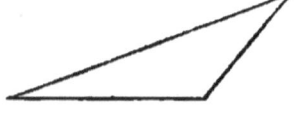

Hypothenuse, from a Greek word, which signifies to *subtend* or *stretch*—a line subtended from the base to the perpendicular.

OBTUSE-ANGLED TRIANGLE.

An *Obtuse-angled Triangle* is one having an obtuse angle.

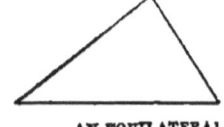

ACUTE-ANGLED TRIANGLE.

An *Acute-angled Triangle* is one having three acute angles.

AN EQUILATERAL TRIANGLE.

An *Equilateral Triangle* has all three of its sides equal.

Equilateral, from the Latin *æquus,* equal, and *lateralis,* from *latus,* side.

---

17. What is a triangle? How many kinds? Describe (or draw) a right-angled triangle. Describe its parts. (Hypothenuse?) An obtuse? Acute?

DEFINITIONS. 21

ISOSCELES TRIANGLE.

An *Isosceles Triangle* has only two of its sides equal.

The term *Isosceles* is from a Greek word, signifying *equal legs;* hence a triangle with two equal legs is called an Isosceles Triangle.

A *Scalene Triangle* is one having no two sides equal.

The term *Scalene* is from the Greek *skalenos*, and signifies *oblique, unequal.* (See obtuse and acute angled.)

18. CIRCLES AND ELLIPSES.

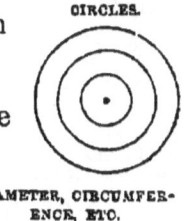

A CIRCLE.

CONCENTRIC CIRCLES.

A *Circle* is a plane figure, bounded by a curve line, every part of which is equally distant from a point within called the center.

Concentric Circles are such as are drawn around a common center.

The *Circumference* of a circle is the curve line which bounds it.

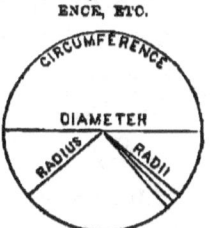

DIAMETER, CIRCUMFERENCE, ETC.

The *Diameter* of a circle is a right line passing through its center, and terminating each way in the circumference.

The *Radius* of a circle is a right line drawn fom its center to any point in the circumference.

The plural of radius is *radii;* and as radii proceed from a common center, light, which proceeds from a luminous point in all directions, is said to *radiate;* and the light thus dispersed is sometimes called *radiations* or *radiance.*

All circles, whether great or small, are supposed to be divided into 360 equal parts, called *degrees;* each degree into 60 equal parts, called *minutes;* and each minute into 60 equal parts, called *seconds.* They are marked respectively thus: Degrees (°), minutes ('), seconds (").

Equilateral? (Derivation?) Isosceles? (Derivation?) Scalene? (Derivation?)
18. What is a circle? Concentric circles? The Circumference? Diameter? Radius? (Plural, &c.?) How all circles divided? (What is a *protractor?*

To save the trouble of dividing a circle into 360°, in order to measure the degrees of an angle, we make use of an instrument called a *Protractor*. It consists of a semicircle of silver or brass, divided into degrees, as represented in the inclosed figure. To measure an angle, as A B C, the straight edge of the protractor is placed upon the line B C, so that the center around which it is drawn will be exactly at the intersection of the lines, or point of the angle, as at B; then the number of degrees included between the lines on the protractor will represent the *quantity* or amount of the angle. From this it will be seen that the amount of the angle does not depend upon the length of the lines which form it, nor upon the magnitude of the circle on which the degrees are marked by which it is measured, but simply upon the width of the opening between the lines, as compared with the whole circumference around the point B. A circle marked off into degrees, minutes, and seconds, is called a *graduated circle*.

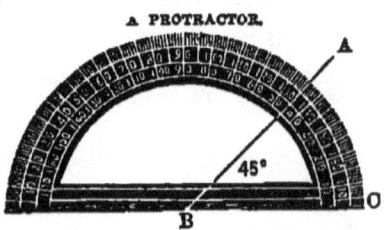

A PROTRACTOR.

Circles are also divided into *Semicircles*, *Quadrants*, *Sextants*, *Signs*, and *Arcs*.

A *Semicircle* is the half of a circle, or 180°.

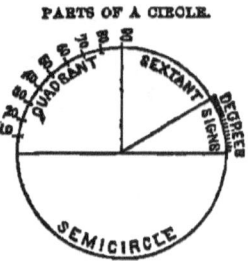

PARTS OF A CIRCLE.

A *Quadrant* is one quarter of a circle, or 90°.

The term *Quadrant* is applied to a nautical instrument, of the form of a quarter of a circle, which is much used by navigators in determining the altitude or apparent hight of the sun, moon, and stars.

A *Sextant* is the sixth part of a circle, and contains 60°.

The word *Sextant* also denotes an instrument similar to a Quadrant, and is used for similar purposes. The main difference is, that one represents 60°, and the other 90°, of a circle. The *Octant*, or eighth part of a circle, is also used for similar purposes.

A *Sign* is the twelfth part of a circle, or 30°.

An *Arc* is any indefinite portion of a circle.

The word *Arc* is from the Latin *arcus*, a bow, vault, or arch. By associating the word *arc* with *arch*, the student may always remember its meaning.

A *Chord* is a right line, joining the extremities of an arc.

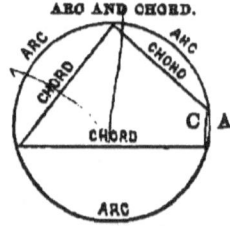

ARC AND CHORD.

The Chord of an Arc is said to be *subtended* (from *sub*, under, and *teno*, to stretch), because it seems stretched under the arc like the string of a bow. In the cut, there are four arcs, and as many chords. The lower arc is a large one, while the arc and chord, A C, are quite small. Still each division of the circle, whether great or small, is an *arc*, and the line joining the extremities of each arc, respectively, is a *chord*.

Describe. A graduated circle ?) What larger divisions of a circle ? What is a semicircle ? A quadrant ? (Note.) A sextant ? (Note.) A sign ? An arc ? (Derivation of term ?) Define a chord. (Why said to be subtended ?)

An *Ellipse* is an oblong figure like an oblique view of a circle, having two points called its *foci*, around which, as centers, the figure is described.

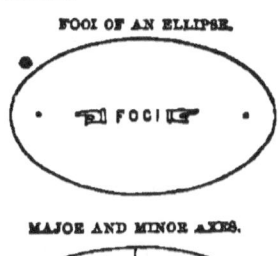

FOCI OF AN ELLIPSE.

Foci is the plural of *focus*.

The longer diameter of an ellipse is called its *Major Axis*, and the shorter its *Minor Axis*.

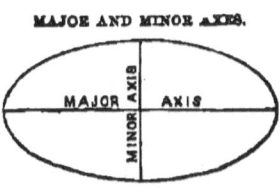

MAJOR AND MINOR AXES.

Axes is the plural of *axis*. The longer is sometimes called the *Transverse*, and the shorter the *Conjugate*, Axis; but major and minor are more simple and perspicuous, and therefore preferable.

The *Eccentricity* of an ellipse is the distance between its center and either focus.

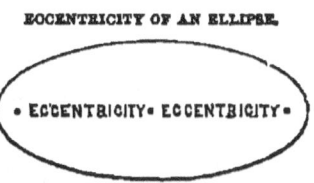

ECCENTRICITY OF AN ELLIPSE.

Eccentric—ex, from, and *centrum, center.* Hence a circle that varies in its distance from the center is *eccentric.* So, also, persons who depart from the usual round of thought and custom are called *eccentric* persons.

19. THE TERRESTRIAL SPHERE.

The *Terrestrial Sphere* is the earth or globe we inhabit.

1. Though the earth is not, strictly speaking, a *sphere*, as that figure is defined (14), but rather an *oblate spheroid* (14), still it is usually called a *sphere* on account of its near approach to that figure, and as a matter of convenience.
2. *Terrestrial*, Latin *terrestris*, from *terra*, the *earth*. "There are also celestial bodies, and bodies terrestrial; but the glory of the celestial is one, and the glory of the terrestrial is another."—1 Cor. xv. 40.

The *Axis* of the earth is the imaginary line about which it makes its daily revolution.

The *Poles* of the earth are the extremities of her axis where they cut or pass through the earth's surface.

The wire upon which an artificial globe turns represents the earth's axis, and the extremities the North and South Poles.

The *Equator* of the earth is an imaginary circle drawn around it, from east to west, at an equal distance from the poles, and dividing it into two equal parts, called Hemispheres.

See illustration, page 18.

An ellipse? Its foci? (Plural and singular?) Major and minor axes? (Singular and plural?) Eccentricity of an ellipse? (Derivation?)

19. The terrestrial sphere? (Is the earth a sphere? Derivation of term terrestrial?) Axis of the earth? Poles? Equator? Latitude? Parallels?

24 ASTRONOMY.

Latitude upon the earth is distance either North or South of the Equator, and is reckoned each way toward the Poles in Degrees, Minutes, and Seconds.

As the distance from the Equator to the Pole cannot be more than a quarter of a circle, or 90°, it is obvious that no place can have more than 90° of latitude; or, in other words, all places upon the earth's surface must be between the Equator and 90° of latitude, either north or south.

Parallels of Latitude are circles either North or South of the Equator, and running parallel to it.

We may imagine any conceivable number of parallels between the Equator and the Poles, though upon most maps and globes they are drawn only once for every ten degrees.

PARALLELS.

The *Tropics* are two parallels of latitude, each 23° 28′ from the Equator.

The Northern is called the *Tropic of Cancer*, and the Southern the *Tropic of Capricorn*.

THE TROPICS AND POLAR CIRCLE.

1. The Tropical Circles are shown at E E in the annexed figure.
2. The sun never shines perpendicularly upon any points on the earth further from the Equator than the Tropics. Between these he seems to travel regularly, leaving the Southern Tropic on the 23d of December, crossing the Equator northward on the 20th of March, reaching the Northern Tropic on the 21st of June, crossing the Equator southward on the 23d of September, and reaching the Southern Tropic again on the 23d of December. In this manner he seems to cross and recross the Equator, and vibrate between the Tropics from year to year. The *cause* of this apparent motion of the sun will be explained hereafter.

The *Polar Circles* are two parallels of latitude, 23° 28′ from the Poles. (See F F in the last cut.)

The *Northern* is called the *Arctic*, and the Southern the *Antarctic*, Circle.

The *Tropics* and *Polar Circles* divide the globe into five principal parts, called Zones, namely, one Torrid, two Temperate, and two Frigid.

A *zone* properly signifies a *girdle;* but the term is here used in an accommodated sense, as only three of these five divisions at all resemble a girdle. The parts cut off by the polar circles are mere convex segments of the earth's surface.

The tropics? Names? Polar circles? Names? Zones? Names? (Are there in reality any frigid *zones?*) Situation of the several zones? Meridians? Longitude on the earth? First meridian? (European and American charts and globes?) How longitude reckoned? Its greatest extent?

The Torrid Zone is situated between the Tropics; the Temperate, between the Tropics and the Polar Circles; and the Frigid, between the Polar Circles and the Poles.

THE FIVE ZONES.

Meridians are imaginary lines drawn from pole to pole over the earth's surface.

Meridians cross the Equator at right angles; and the plane of any two Meridians directly opposite each other would divide the earth into Eastern and Western Hemispheres, as the Equator divides it into Northern and Southern. We may imagine Meridians to pass through every conceivable point upon the earth's surface. They meet at the Poles, and are furthest apart at the Equator.

MERIDIANS.

Longitude upon the earth is distance either East or West of any given meridian.

A degree of longitude at the Equator comprises about 69½ miles, but is less and less as the meridians approach the Poles, at which points it is nothing. A degree of latitude is about 69½ miles on all parts of the globe.

The *First Meridian* is that from which the reckoning of Longitude is commenced.

On European charts and globes, longitude is usually reckoned from the Royal Observatory at Greenwich, near London; but in this country it is often reckoned from the Meridian of Washington. It would be better for science, however, if all nations reckoned longitude from the same Meridian, and all charts and globes were constructed accordingly.

As Longitude is reckoned both East and West, the greatest longitude that any place can have is 180°.

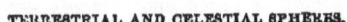
TERRESTRIAL AND CELESTIAL SPHERES.
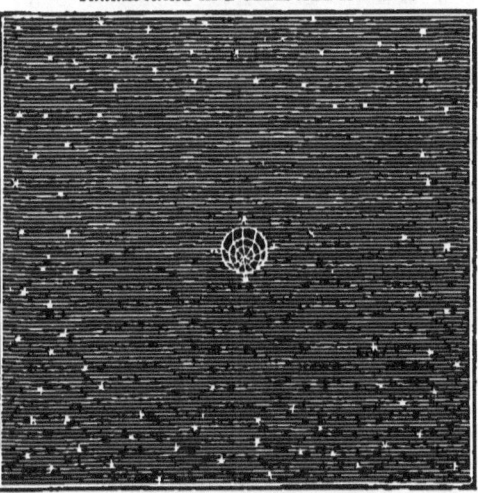

20. THE CELESTIAL SPHERE.

The *Celestial Sphere* is the apparent concave surface of the heavens, surrounding the earth in all directions.

The relation of the Terrestrial to the Celestial Sphere may be understood by the above diagram, in which the stars surround the earth in all directions, as they seem to fill the whole celestial vault.

2

The *Axis of the Heavens* is the axis of the earth produced or extended both ways to the concave surface of the heavens.

The *Equator of the heavens*, or *Equinoctial*, is the plane of the Earth's equator extended to the starry heavens.

EQUATOR OF THE HEAVENS, OR EQUINOCTIAL.

Declination is distance either north or south of the Equinoctial.

Declination is to the heavens precisely what latitude is upon the earth. It is reckoned from the celestial equator, both North and South, to 90°, or to the poles of the heavens. Celestial Latitude can be explained better hereafter, and so with the terms *Ecliptic, Zodiac,* &c.

Right Ascension is distance east of a given point, and is reckoned on the Equinoctial quite around the heavens.

In one respect, Right Ascension in the heavens is like longitude on the earth: they are both reckoned upon the equators of their respective spheres. But while longitude is reckoned both east and west of the first meridian, and can only amount to 180°, Right Ascension is reckoned *only eastward*, and consequently may amount to 360°, or the whole circle of the heavens. The principal difference between Right Ascension and Celestial Longitude is, that the former is reckoned on the Equinoctial, and the latter on the Ecliptic.

The *Sensible Horizon* is that circle which terminates our view, or where the earth and sky seem to meet.

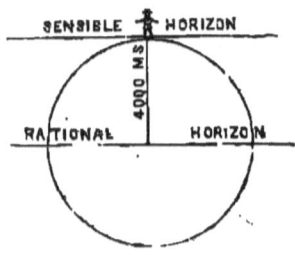

The *Rational Horizon* is an imaginary plane, below the visible horizon, and parallel to it, which, passing through the earth's center, divides it into upper and lower hemispheres.

1. These hemispheres are distinguished as *upper* and *lower* with reference to the observer only.

20. Celestial sphere? (Relation to terrestrial?) Axis of the heavens? Equator of the heavens? Declination? (How illustrated by terrestrial latitude? How reckoned? Its limits?) Right ascension? (How resembles longitude? What difference?) Sensible horizon? Rational? Explain by

DEFINITIONS. 27

2. The *sensible horizon* is half the diameter of the earth, or about 4,000 miles from the *rational*; and yet so distant are the stars, that both these planes seem to cut the celestial arch at the same point; and we see the same hemisphere of stars above the sensible horizon of any place that we should if the upper half of the earth were removed, and we stood on the rational horizon of that place.

The *Poles of the Horizon* are two opposite points—one directly above, and the other directly beneath, us. The first is called the *Zenith*, and the latter the *Nadir*.

The points *Up* and *Down*, *East* and *West*, are not positive and permanent directions, but merely relative.

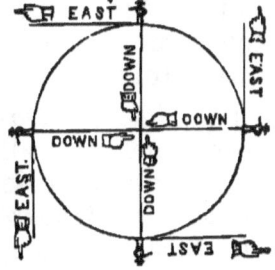

UP AND DOWN, AND EAST AND WEST.

1. As the earth is a sphere, inhabited on all sides, the Zenith point is merely *opposite its center*, and the Nadir *toward its center*. So with the directions *Up* and *Down*: one is *from* the center, and the other *toward* it; and the same direction which is *up* to one, is *down* to another. This fact should not merely be acknowledged, but should be dwelt upon until the mind has become familiarized to the conception of it, and divested, as far as possible, of the notion of an absolute up and down in space. We should remember that we are bound to the earth's surface by attraction, as so many needles would be bound to the surface of a spherical loadstone.

2. *East* and *West* also are not absolute, but merely relative directions. East is that direction in which the sun appears to rise, and West is the opposite direction; and yet, so far as absolute direction is concerned, what is East to one, as to the observer at A, is West to B, and so with C and D. And as the earth revolves upon its axis every twenty-four hours, it is obvious that East and West upon its surface must, in that time, change to every point in the whole circle of the heavens. The same is true of the Zenith and Nadir, or of up and down.

Space, in Astronomy, is that boundless interval or void in which the earth and the heavenly bodies are situated, and extending infinitely beyond them all, in every direction.

Space has no limits—or, in other words, is *boundless*, or *infinite*. Suppose six persons were to start from as many different points upon the earth's surface—as, for instance, one from each pole, and one from each of the positions occupied by observers in the next figure. Let them ascend or diverge from the earth in straight lines, perpendicularly, to its surface, and though they were to proceed onward, separating from each other, with the speed of lightning, for millions of ages, none of these celestial voyagers would find an end to space, or any effectual barrier to hinder their advancement. Should they chance to meet another world in the line of their flight, it would soon be passed, like a ship met by a mariner upon the ocean, and beyond it *space* would still invite them onward to explore its immeasurable depths. And thus they might go on *forever*, without changing their position in respect to the *center* or *boundaries* of immensity; for as *eternity* has no beginning, middle, or end, so *space* is without center or circumference—an ethereal ocean, without bottom or shore.

---

diagram. Poles of the horizon? Names? Up and down—positive or relative points? (Illustrate by diagram; also east and west.) Term space in astronomy? (Has it any limits? Illustration.)

21. First Grand Divisions of the Universe.

The visible universe may be considered under two grand divisions—viz., the SOLAR SYSTEM and the SIDEREAL HEAVENS.

The *Solar System* consists of the sun and all the planets and comets that revolve around him.

The *Sidereal Heavens* include all those bodies that lie *around* and *beyond* the Solar System, in the region of the Fixed Stars.

THE SOLAR SYSTEM.

SOLAR SYSTEM AND SIDEREAL HEAVENS.

1. The word *Sidereal* is from the Latin *sideralis*, and signifies *pertaining to the stars*. The Sidereal Heavens are, therefore, the heavens of the fixed stars.
2. The relation of the Solar System to the Sidereal Heavens is shown in the annexed cut, where the sun appears only as a *star*, at a distance from all others, and surrounded by his own retinue of worlds. The Solar System is drawn upon a small scale, and the Sidereal Heavens are seen around and at a distance from it in every direction.

In considering the general subject of Astronomy, we shall proceed according to the foregoing classification, treating first of the SOLAR SYSTEM, and, secondly, of the SIDEREAL HEAVENS.

21. How visible universe divided? Define each? (Derivation of term sidereal? Relation of solar system to the sidereal heavens? Illustrate by drawing.) Of which division does the author first treat?

PART I.

THE SOLAR SYSTEM.

CHAPTER I.

THE PRIMARY PLANETS.

22. The *Solar System* derives its name from the Latin term *sol*, the *sun*. It signifies, therefore, the System of the Sun. It includes that great luminary, and all the planets and comets that revolve around him.

23. The *Sun* is the fixed center of the system, around which all the solar bodies revolve, and from which they receive their light and heat.

24. The *Planets* are those spherical bodies or worlds that revolve statedly around the sun, and receive their light and heat from him.

The term *planet* signifies a *wanderer*, and was applied to the solar bodies because they seemed to move or wander about among the stars.

The *Orbit* of a planet is the path it pursues in its revolution around the sun.

25. The planets are divided into *Primary* and *Secondary* planets.

The *Primary Planets* are those larger bodies of the system that revolve around the sun only, as their center of motion.

The *Secondary Planets* are a smaller class of bodies,

---

22. Of what does Part II. treat? What meant by the Solar System? Includes what?
23. What is the sun?
24. Describe the planets. (The term?) The orbit of a planet?
25. How planets divided? Describe each. (What other names for secondaries?)

30　　　　　　　　　ASTRONOMY.

that revolve not only around the sun, but also around the primary planets, as their attendants, or moons.

<small>The secondary planets are also called *Moons* or *Satellites*. A satellite is a *follower* or attendant upon another.</small>

VIEW OF THE SOLAR SYSTEM.

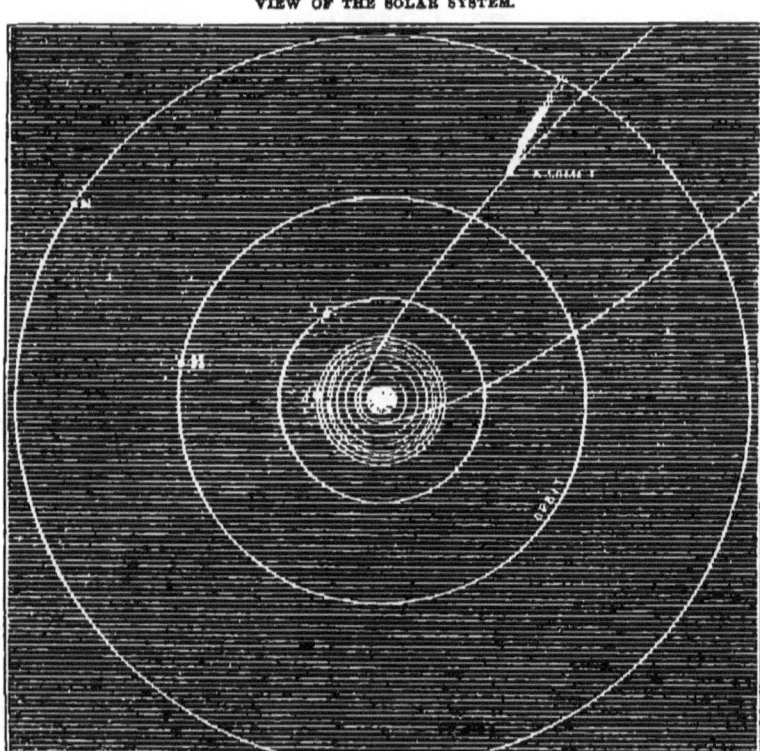

<small>In this cut, the sun may be seen in the center. The white circles are the *Orbits* of the primary planets. The *planets* may be seen in those orbits at various distances from the sun. The numerous orbits so close together are those of the *Asteroids*. The secondary planets may be seen near their respective primaries, revolving around them, while they all go on together around the sun. On the right is seen a *Comet* plunging into the system, with his long and fiery train. His orbit is seen to be very elliptical. All these bodies are opake, the sun excepted. Even the blazing comet shines only by reflection.</small>

26. The planets are again divided into *Interior* and *Exterior* planets.

The *Interior Planets* are those whose orbits lie *within* the orbit of the earth, or between it and the sun.

<small>26. What meant by interior and exterior planets? (Why not inferior and superior?)</small>

The *Exterior Planets* are those whose orbits lie *without* the orbit of the earth.

Some Astronomers speak of these two classes respectively as *Inferior* and *Superior*. The reason seems to be, that as those nearer the sun than the earth are *lower* than she is—that is, nearer the great center of the system—they are, in this respect, *inferior* to her; while, on the other hand, those that are *above*, or beyond her, are her *superiors*. But as the distinction is founded upon, and is intended to denote, the *position* of the planets with respect to the earth's orbit, it is obvious that *interior* and *exterior* are the more appropriate terms. It seems hardly allowable to call the Asteroids superior planets, and Mercury and Venus, which are much larger, inferior.

27. Eighteen of the smaller primary planets are called *Asteroids*.

Asteroid signifies *star-like*, and is applied to these small planets because of their comparative minuteness. They are never seen except through telescopes, and through ordinary instruments are not always readily distinguished from the fixed stars.

28. *Comets* are a singular class of objects, belonging to the solar system, distinguished for their long trains of light, their various shapes, and the great eccentricity of their orbits.

NUMBER AND NAMES OF THE PRIMARY PLANETS.

29. The *Primary Planets* are thirty-five in number. They are denoted, in astronomical works, by certain characters, or symbols; and as the names of the planets are mostly derived from Mythology, their symbols generally relate to the imaginary divinity after whom the planet is named. The names of the planets and their symbols are as follows:

Mercury	☿	Irene,	
Venus	♀	Eunomia	
Earth	⊕	Juno	
Mars	♂	Ceres	
Flora		Pallas	
Clio		Hygeia	
Vesta		Melpomene	
Iris		Misillia	
Metis		Anonymous	
Hebe		Jupiter	♃
Parthenope		Saturn	♄
Egeria*		Uranus	♅
Astræa		Neptune	♆

Several, recently discovered, are not in this table.

27. What are the asteroids? How many? (Term? Are they visible to the naked eye?)
28. What are comets? (Describe the preceding cut. Where sun? Prim-

1. The planets are placed in the order of their distances, respectively, beginning at the sun.
2. We have not in every case been able to procure the astronomical symbol. This accounts for the blanks opposite several of the names.
3. The names of the eighteen asteroids are included in braces.

MYTHOLOGICAL HISTORY AND SYMBOLS.

30. MERCURY was the messenger of the gods, and the patron of thieves and dishonest persons. His symbol denotes his *caduceus*, or *rod*, with serpents twined around it (☿).*

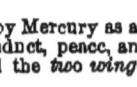
ROD OF MERCURY

1. Mercury was represented as very eloquent, and skillful in interpreting and explaining—as the god of rhetoricians and orators. Hence, when Paul and Barnabas visited Lystra, addressed the people, and wrought a miracle, they said, "The gods have come down to us in the likeness of men. And they called Barnabas Jupiter, and Paul *Mercurius, because he was the chief speaker.*"
2. "The *caduceus* of Mercury was a sort of wand or scepter, borne by Mercury as an ensign of quality and office. On medals, it is a symbol of good conduct, peace, and prosperity. The *rod* represents *power;* the *serpents, wisdom;* and the *two wings, diligence and activity.*"—ENCYCLOPÆDIA.
3. The original form of this sign may be understood by the preceding cut, to which the present astronomical symbol (☿) bears but a slight resemblance.

31. VENUS was the goddess of love and beauty, and her sign is an ancient *mirror* or *looking-glass* (♀), which she is represented as carrying in her hand.

MIRROR OF VENUS.

Anciently, mirrors were made of *brass* or *silver*, highly polished, so as to reflect the image of whatever was brought before them. Hence it is said in the Book of Exodus, written fifteen centuries before Christ, that Moses "made the laver of *brass,* and the foot of it of *brass,* of the *looking-glasses* of the women," &c. For convenience, the ancient mirrors had a handle attached, as represented in the cut, which very much resembles the sign of the planet.

32. THE EARTH (called by the Greeks *Ge*, and by the Latins *Terra*) has two symbols—one representing a sphere and its equator (⊖), and the other (⊕) the four quarters of the globe.

---

* All these symbols should be drawn in rotation upon the Blackboard, during recitation, by the Teacher, or some member of the class. It will be well, therefore, for the student to observe each sign carefully, that he may be prepared to *draw* and *explain* it, if called upon.

---

aries? Secondaries? Asteroids? Orbits? Comet and orbit? Which self-luminous, and which opake?)
29. How many primary planets? How represented in astronomical works? Origin of names and symbols? Repeat names. Draw symbols on blackboard. (In what order arranged? How asteroids designated?)
30. Who was Mercury, in Mythology, and what does his symbol denote? (How was he represented? What Scriptural allusion? Describe his caduceus. The meaning of its parts?)

33. MARS was the god of war, and his sign (♂) represents an ancient *shield* or *buckler*, crossed by a *spear*.

SPEAR AND SHIELD OF MARS.

Gunpowder was not known to the ancients, consequently they had no pistols, muskets, or cannon. They fought with short swords and spears, and defended themselves with the *shield*, carried on the left arm. A shield and spear were, therefore, very appropriate emblems of war. The original form of the sign of Mars is presented in the cut.

34. FLORA was the "queen of all the flowers," and her symbol (⚹) is a *flower*, the "Rose of England."

35. CLIO was one of the *Muses*. Her sign (⚹) is a *star*, with a *sprig of laurel* over it.

36. VESTA was the goddess of *fire*, and her sign (⚶) is an *altar*, with a *fire* blazing upon it.

37. IRIS was the beautiful waiting-maid of Juno, the queen of heaven. Her symbol (⚵) is composed of a semicircle, representing the *rainbow*, with an interior *star*, and a base line for the horizon.

"As an attendant upon Juno," says Prof. Hind, "the name was not inappropriate at the time of discovery, when Juno was traversing the 18th hour of right ascension, and was followed by Iris in the 19th."

38. METIS was the first wife of Jupiter, and the goddess of prudence and sagacity. Her symbol (⚴) is an *eye* (denoting wisdom) and a *star*.

39. HEBE presided over children and youth, and was cup-bearer to Jupiter. Her sign (⚳) is a *cup*.

Hebe was celebrated for her beauty, but happening one day to stumble and spill the nectar, as she was serving Jupiter, she was turned into an *hostler*, and doomed to harness and drive the peacocks of the queen of heaven.

40. PARTHENOPE was one of the three *Syrens*—a sea nymph of rare beauty. They were all admirable *singers;* hence a *lyre* (⚷) is her appropriate sign.

1. The three Syrens—Parthenope, Ligeia, and Leucosia—were represented as dwell-

31. Venus and symbol? (Ancient mirrors? Scripture allusion?)
32. The Earth—ancient name and symbols?
33. Mars and symbol? (Ancient mode of warfare?)
34. Flora and sign?
35. Clio and symbol?
36. Vesta and her symbol?
37. Iris and her sign? (Prof. Hind's remark?)
38. Metis and her sign?
39. Hebe and her sign? (Incident mentioned in note?)
40. Parthenope and sign? (What said of the Syrens? Of the appropriateness of the name?)

ing upon the coast of Sicily, and luring mariners upon the rocks of destruction by their enchanting songs. Hence whatever tends to entice or seduce to ruin is often called a "syren song."

2. As this planet was discovered at the Naples Observatory, in Italy, it was quite appropriate to name it after one of the Syrens, that Mythology located on the coast of a neighboring island.

41. EGERIA was the counselor of Numa Pompilius. Symbol not yet agreed upon by astronomers.

42. ASTRÆA was the goddess of *Justice*, and her sign (♎) is a balance.

Mythology teaches that Justice left heaven, during the golden age, to reside on the earth; but becoming weary with the iniquities of men, she returned to heaven, and commenced a constellation of stars. The constellations Virgo and Libra in the zodiac are representations of Astræa and her golden scales. So the female figure, holding a pair of *scales*, in the coat of arms of several of the United States, is a representation of Astræa, and denotes *Justice*.

43. IRENE was one of the *Seasons*. The planet was so named by Sir John Herschel, in honor of the peace prevailing in Europe at the time of its discovery (May, 1851). Its symbol (☧) is a *dove*, with an *olive branch* in her mouth, and a *star* upon her head.

44. EUNOMIA was another of the *Seasons*—a sister of Irene. (Symbol not ascertained.)

45. JUNO was the reputed queen of heaven, and her sign (☿) is an ancient *mirror*, crowned with a star—an emblem of beauty and power.

46. CERES was the goddess of *grain* and *harvests*, and her sign (⚳) is a *sickle*.

47. PALLAS (or Minerva) was the goddess of *wisdom* and of *war*. Her symbol (⚴) is *the head of a spear*.

1. The ancient *Palladium* was an image of Pallas, preserved in the castle of the city of Troy; for while the castle of the city of Minerva was building, they say this image fell from heaven into it, before it was covered with a roof.—*Tooke's Pantheon.*

2. To a similar fable, respecting an image falling from heaven, the Apostle Paul alludes, Acts xix. 35:—"Ye men of Ephesus, what man is there that knoweth not how that the city of Ephesus is a worshiper of the great goddess Diana, and of the *image* which fell down from Jupiter?"

---

41. Egeria and her symbol?
42. Astræa and sign? (Mythological legend? Virgo and Libra? Where else found?)
43. Irene—by whom named, and why? Symbol?
44. Eunomia and symbol?
45. Juno and symbol?
46. Ceres and her symbol?
47. Pallas and her symbol? (Ancient *Palladium?* Reputed origin? Scriptural allusion to it?)

48. HYGEIA was the goddess of *health*, and the daughter of Esculapius, the father of the healing art. (Symbol not ascertained.)

<small>Our modern word *Hygeian*, which signifies the laws of health, &c., is derived from the goddess Hygeia.</small>

49. JUPITER was the reputed father of the gods—the king of heaven. His symbol (♃) was originally the Greek letter ζ, *zeta* (the same as our Z)—the initial of the Greek word *Zeus*, the name for Jupiter.

50. SATURN—called by the Greeks *Chronos*—presided over *time* and *chronology*. His sign (♄) represents a *scythe*.

SATURN, OR CHRONOS.

<small>1. Saturn was represented in Mythology as an old man, with wings, bald excepting a forelock, with a scythe in one hand, and an hourglass in the other. The same figure is now used to represent time.
2. Our modern word *chronology*, from *chronos*, time, and *logos*, discourse, signifies the science of keeping time, dates, &c.</small>

51. URANUS was the father of Saturn, and presided over astronomy. The symbol of this planet (♅) consists of the letter H, with a planet suspended from the cross-bar, in honor of Sir William Herschel, its discoverer.

<small>This planet is popularly known by the name of Herschel, but astronomers now almost universally call it Uranus. It bears this name in the British *Nautical Almanac* for 1851, with the full consent of Sir John Herschel, the son of the great discoverer. It was first called *Georgium Sidus*, by Dr. Herschel, in honor of his royal patron. George III.</small>

52. NEPTUNE was the god of the seas, but the symbol of the planet (♆) is composed of an L and a V united, with a planet suspended from the hair-line of the V, in honor of Le Verrier, its discoverer.

<small>This planet was first called *Le Verrier*, but is more generally known by the name of Neptune.</small>

53. The MOON was called *Luna* by the Romans, and

<small>48. Hygeia and symbol? (Term *hygeian*?)
49. Jupiter and his symbol?
50. Saturn? Greek name? Symbol? (How represented in Mythology Word *chronology*?)
51. Uranus and symbol? (What other names, and why?)
52. Neptune and his symbol? (Former name?)</small>

Selene by the Greeks. She is known by various symbols, according as she is new, half-grown, or full, thus: ☽, ☽, ○.

<small>1. From *Luna* we have our modern terms *lunar* and *lunacy;* the former of which signifies pertaining to the moon, and the latter a disease anciently supposed to be caused by the moon.
2. *Selene,* in Mythology, was the daughter of *Helios,* the Sun. Our English word *selenography*—a description of the moon's surface—is from *Selene,* her ancient name, and *grapho,* to describe.</small>

54. The Sun—called *Sol* by the Romans, and *Helios* by the Greeks—is represented by a *shield* or buckler, thus: ☉, ☉, ☉. As the large and polished bucklers of the ancients dazzled the eyes of their enemies, this instrument was selected as an appropriate emblem of the sun.

DISTANCES OF THE PLANETS.

55. The *distances* of the planets from the sun, commencing with Mercury, and proceeding outward, are as follows; viz.:

Planet	Round		Exact
Mercury	37 millions,	or	36,890,000
Venus	69	"	68,770,000
Earth	95	"	95,298,260
Mars	145	"	145,205,000
The Asteroids, from	210	" to	300,000,000
Jupiter	496	" or	495,817,000
Saturn	909	"	909,028,000
Uranus	1,800	"	1,829,071,000
Neptune	2,862	"	2,862,457,000

<small>1. The first column of round numbers only should be committed to memory by the student. These should be well fixed in the mind, as it will greatly facilitate the progress of the student hereafter. The family of Asteroids being less important, their distances need not be learned in detail. The following table shows the distances of the several Asteroids from the sun:</small>

Flora	209,826,000	Astræa	245,622,000
Clio	222,378,000	Irene	246,079,000
Vesta	225,000,000	Eunomia	252,300,000
Iris	227,334,000	Juno	254,312,000
Metis	227,387,000	Ceres	263,718,000
Hebe	231,089,000	Pallas	264,256,000
Parthenope	233,611,000	Hygeia	300,322,000
Egeria	244,940,000		

<small>53. The Moon—Latin and Greek names? Symbols? (Words *lunar* and *lunacy?* Who was *Selene* in Mythology? Selenography? Derivation?)
54. The Sun—Latin and Greek names? Symbol, and why?
55. Rehearse, in round numbers, the *distances* of the planets from the (Substance of note 1st? Object of note 2d? Note 3d? Note 4th?)</small>

DISTANCES OF THE PLANETS. 37

2. The comparative distances of the planets are represented in the cut, page 15, and also in the following:

COMPARATIVE DISTANCES OF THE PLANETS.

3. To assist his conception of these vast distances, the student may imagine a railroad laid down from the sun to the orbit of Neptune. Now if the train proceed from the sun at the rate of thirty miles an hour, without intermission, it will reach *Mercury* in 152 years; the *Earth* in 361 years; *Jupiter* in 1,864 years; *Saturn* in 3,493 years; *Herschel* in 6,933; and *Neptune* in 10,800 years! Such a journey would be equal to riding 900,000 times across the continent, from Boston to Oregon!

4. It is now about 5,850 years since the creation of the world. Had a train of cars started from the sun at that time toward the orbit of Neptune, and traveled day and night ever since, it would still be 284 millions of miles within the orbit of Herschel—about where the head of the locomotive stands, as shown in the cut! To reach even that planet would require over 1,000 years longer; and to arrive at Neptune, nearly 6,000 years to come! Such is the vast area embraced within the orbits of the planets, and the spaces over which the sunlight travels, to warm and enlighten its attendant worlds.

56. The apparent magnitude of the heavenly bodies depends much upon the *distance* from which they are viewed; the magnitude increasing as the distance is diminished, and diminishing as the distance is increased.

NEAR AND REMOTE VIEWS OF THE SAME OBJECT.

Let A represent the position of an observer upon the earth, to whom the sun appears 32', or about half a degree in diameter. Now it is obvious that if the observer advance to B (half way), the object will fill an angle in his eye *twice as large* as it filled when viewed from A. Again: if he recede from A to C, the object will appear but *half* as large. Hence the rule, that the apparent magnitude is increased as the distance is diminished, and diminished as the distance is increased.

57. Could a beholder leave the earth, and, descending toward the sun, station himself upon Mercury, he would find the apparent magnitude of the sun vastly increased. Should he then return, and pass outward to Mars or Jupiter, he would observe a corresponding diminution in the sun's magnitude, in proportion as the distance was increased. Hence the apparent magnitude must vary

---

56. How apparent magnitudes of heavenly bodies modified? (Illustrate by diagram.)
57. Suppose a person to go to Mercury—what effect upon apparent size of the sun?

exceedingly, as viewed from different points in the solar system.

THE SUN, AS SEEN FROM THE DIFFERENT PLANETS.

The above cut represents the relative apparent magnitude of the sun, as seen from the different planets. In angular measurements, its diameter would be as follows:

From Mercury	. 82½′	From Jupiter	. .	6′
" Venus	. . 44½′	" Saturn	. . .	3⅕′
" Earth	. . 32′	" Uranus	. .	1⅔′
" Mars	. . 21′	" Neptune	. .	50″
The Asteroids, say 12′				

Let us continue our imaginary journey outward, beyond Neptune, toward the fixed stars, and in a short time the glorious sun, so resplendent and dazzling to our view, will appear only as a sparkling *star;* and the fixed stars will expand to view as we approach them, till they assume all the magnitude and splendor of the sun himself.

LIGHT AND HEAT OF THE PLANETS.

58. As the distances of the planets, respectively, affect the apparent magnitude of the sun, as viewed from their surfaces, so it must affect the relative amount of *light* and *heat* which they respectively receive from this great luminary.

59. The amount of light and heat received from the sun, by the several planets, is in inverse proportion to the square of their respective distances.

---

58. What effect has the *distances* of the planets from the sun, respectively upon their relative light and heat?
59. What rule governs the diffusion of light? (Illustrate by a diagram.)

PHILOSOPHY OF THE DIFFUSION OF LIGHT.

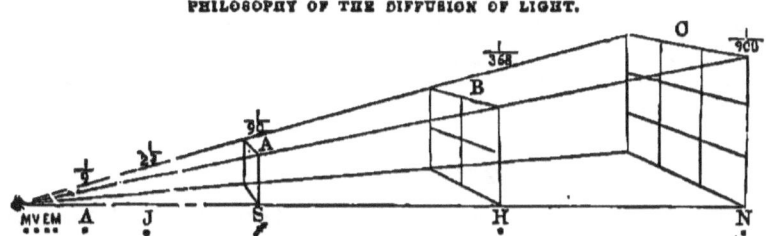

1. Here the light is seen passing in right lines, from the sun on the left toward the several planets on the right. It is also shown that the surfaces A, B, and C receive equal quantities of light, though B is four times, and C nine times, as large as A; and as the light falling upon A is spread over four times as much surface at B, and nine times as much at C, it follows that it is only one-ninth as intense at C, and one-fourth at B, as it is at A. Hence the rule, that *the light and heat of the planets is, inversely, as the squares of their respective distances.*

2. The student may not exactly understand this last statement. The square of any number is its product, when multiplied by itself. Now suppose we call the distances A, B, and C 1, 2, and 3 miles. Then the square of 1 is 1; the square of 2 is 4; and the square of 3 is 9. The light and heat, then, would be in *inverse* proportion at these three points, as 1, 4, and 9; that is, four times less at B than at A, and nine times less at C. These amounts we should state as 1, ¼, and ⅑.

60. The intensity of light and heat received upon the several planets varies, according to their respective distances, from 6½ times as much as our globe to $\frac{1}{900}$th part as much.

1. The comparative light and heat of the planets—the earth being 1—is as follows:

Mercury	6¼	Jupiter	$\frac{1}{27}$
Venus	2	Saturn	$\frac{1}{90}$
The Earth	1	Uranus	$\frac{1}{360}$
Mars	½	Neptune	$\frac{1}{900}$
The Asteroids	¼		

2. From this table it appears that Mercury has 6¼ times as much light and heat as our globe, Uranus only $\frac{1}{362}$, and Neptune only $\frac{1}{900}$th part as much. Now if the average temperature of the earth is 50 degrees, the average temperature of Mercury would be 325 degrees; and as water boils at 212, the temperature of Mercury must be 113 degrees above that of boiling water. Venus would have an average temperature of 100 degrees, which would be twice that of the earth. On the other hand, Jupiter, Saturn, Uranus, and Neptune, seem doomed to the rigors of perpetual winter. And what conception can we form of a region 900 times as cold as our globe! Surely,

> "Who there inhabit must have other powers,
> Juices, and veins, and sense, and life, than ours;
> One moment's cold, like theirs, would pierce the bone,
> Freeze the heart's blood, and turn us all to stone!"

3. It is not certain, however, that the heat is proportionate to the light received by

---

60. Between what limits does the light and heat of the several planets vary? (What would that be for Mercury? For Venus? How with the exterior planets? Poetry? Is it certain that the heat of the planets is in exact proportion to the light they respectively receive? Why not?)

the respective planets, as various local causes may conspire to modify either extreme of the high or low temperatures. For instance, Mercury may have an atmosphere that arrests the light, and screens the body of the planet from the insupportable rays of the sun; while the atmospheres of Saturn, Herschel, &c., may act as a refracting medium to gather the light for a great distance around them, and concentrate it upon their otherwise cold and dark bosoms.

MAGNITUDE OF THE PLANETS.

61. The planets vary as much in their respective *magnitudes*, as in their distances. Their several diameters, so far as known, are as follows:

Mercury	2,950	Astræa	—
Venus	7,900	Irene	—
Earth	7,912	Eunomia	—
Mars	4,500	Juno	1,400
Flora	—	Ceres	163
Clio	—	Pallas	770
Vesta	295	Hygeia	—
Iris	—	Jupiter	88,790
Metis	—	Saturn	79,300
Hebe	—	Uranus	35,000
Parthenope	—	Neptune	31,000
Egeria	—		

1. The *asteroids* are so small and so remote, that measurements of their exact diameters are obtained with great difficulty; hence the numerous blanks in the above table. And even when diameters are given, they are somewhat doubtful.
2. In the case of the other planets, we have given their *mean* or *average* diameters, according to the best authorities. As most of them are more or less oblate, their polar diameters are *less*, and their equatorial *more*, than the amount given in the table.

62. The magnitude of the principal planets, as compared with the earth, is as follows:—Mercury, $\frac{1}{19}$ as large; Venus, $\frac{9}{10}$; Jupiter, 1,400 times as large; Saturn, 1,000 times; Uranus, 90 times; and Neptune, 60.

1. The magnitudes of spherical bodies are to each other as the cubes of their diameters. Thus, $7912 \times 7912 \times 7912 = 495,289,174,428$, the cube of the earth's diameter; and $2950 \times 2950 \times 2950 = 25,672,375,000$, the cube of the diameter of Mercury. Divide the former by the latter, and we have 19 and a fraction as the number of times the bulk of Mercury is contained in the earth.

---

61. State the diameters of the several planets? (Why blanks in the table? What diameters are given—polar, equatorial, or neither?)
62. Give the magnitude of the principal planets, as compared with the earth. (How ascertain relative magnitudes? How possible that a mere *star* can be such an immense world?)

DENSITY. 41

COMPARATIVE MAGNITUDE OF THE SUN AND PLANETS.

2. It may seem almost incredible that what appear only as small stars in the heavens should be larger than the mighty globe upon which we dwell. But when we consider their immense *distance*, and the effect this must have upon their apparent magnitude, as illustrated at 55, it is evident that the planets could not be seen at all were they not very large bodies. The above cut will give some idea of the magnitude of the several planets, as compared with each other, and also with the sun.

63. The *Sun* is 1,400,000 times as large as our globe, and 500 times as large as all the other bodies of the solar system put together. It would take one hundred and twelve such worlds as our earth, if laid side by side, to reach across his vast diameter.

DENSITY.

64. The planets differ greatly in their *density*, or in the compactness of the substances of which they are composed. Mercury is about three times as dense as our globe, or equal to *lead*. Venus and Mars are about the same as the earth; while Jupiter and Uranus are only $\frac{1}{4}$th as dense, or about equal to water. Saturn has only $\frac{1}{10}$th the density of our globe, answering pretty nearly to *cork*

---

63. State the magnitude of the sun as compared with the earth. With the rest of the system. Illustration?
64. What meant by *density?* Do the planets differ in this respect? State and illustrate. (How *masses* of planets ascertained? How with Mercury?)

The *masses* of the planets are determined by the revolution of their respective satellites; but as Mercury has no satellite, the determination of his mass and density becomes a very difficult and uncertain matter. "But it fortunately happens," says Prof. Hind, "that we have a curious method of approximating to this element, viz., by the perturbations produced by the planet in the movements of a comet known as Euchls, which revolves around the sun in little more than three years, and occasionally approaches very near Mercury, &c. From computations based upon these perturbations, Prof. H. concludes that Mercury is only about $\frac{1\cdot 3}{1\cdot 0\cdot 1}$ more dense than our globe—a result widely different from that arrived at by his predecessors.

GRAVITATION.

65. *Attraction*, or *Gravitation*, is the tendency of bodies toward each other. It is that tendency which causes bodies raised from the earth, and left without support, to fall to its surface.

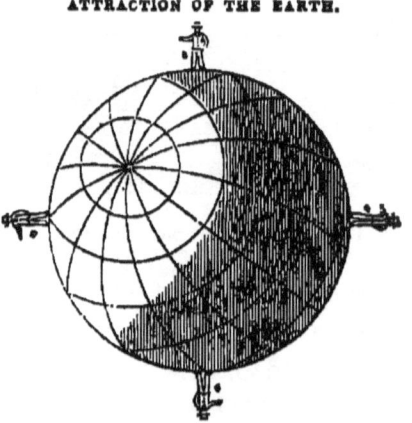

ATTRACTION OF THE EARTH.

All substances fall toward the earth's center from every part of the globe, as a spherical loadstone would attract particles of steel to its surface in every direction. Hence when these four men, standing on different sides of the globe, drop each a stone, they all fall toward the same point, because the earth attracts them all to herself.

66. Gravitation is what constitutes the *weight* of bodies, and depends upon the quantity of matter in the bodies attracting, and their distances from each other.

The reason why a cubic foot of *cork* weighs much less than the same bulk of *lead*, is that being less dense, it contains much less matter to be attracted.

67. From the above law of attraction, it follows that large bodies attract much more strongly than small ones, provided their densities are equal, and their distances the same; and as the force of attraction constitutes the *weight* of a body, it follows that a body weighing a given number of pounds on the earth, would weigh much more on Jupiter or Saturn, and much less on Mercury or the Asteroids.

---

65. Define *gravitation*. (What illustration given?)
66. What relation has gravitation to *weight*? Upon what does it depend for its degree of force? (Why is a cubic foot of *cork* lighter than the same bulk of *lead*?)
67. What effect have the *bulk* and *density* of the planets upon the *weight* of bodies on their surfaces? (State comparative weights. - Illustration? Why not attractive force or weight in exact proportion to bulk? How must bodies be weighed to ascertain difference, and why?)

GRAVITATION. 43

1. The following table shows the relative attractive force of the sun and planets. A body weighing one pound on the earth, would weigh,

	lb. oz.		lb. oz.
On Mercury	1 1¼	On Saturn	1 5¼
" Venus	0 15	" Uranus	0 12¼
" Mars	0 8	" Neptune	unknown.
" Jupiter	2 8	" The Sun	28 5¾

2. A person weighing 150 lbs. on the earth would consequently weigh but 74 lbs. upon Mars; while upon Jupiter, his weight would be 375 lbs.; and upon the sun, 4,250 lbs. The attractive force of the Asteroids is so slight, that, if a man of ordinary muscular strength were transported to one of them, he might probably lift a hogshead of lead from its surface without difficulty.

3. But the learner will notice that the attractive force, as shown in the above table, is not in strict proportion to the *bulk* of the planets respectively. This difference will be accounted for by considering the difference in their *density* (63). From the principles there laid down, it will be seen at once, that though one planet be as large again as another, still, if it were but half as dense, it would contain no more matter than the smaller one, and their attractive force would be equal. If Jupiter, for instance, were as dense as the earth, his attractive force would be four times what it now is; and if the density of all the solar bodies were precisely the same, their attractive force, or the weight of bodies on their surfaces, would be in exact proportion to their bulk.

4. It must be remembered, however, that if a body were actually weighed upon the surface of each planet, by *scales*, it would weigh the same on all, because the force of attraction upon the *weights* would be just equal to that of the body to be weighed, whether it were more or less. With a steelyard it would be the same. A spring and hook, therefore, is the only instrument with which we could weigh objects accurately on the different planets.

68. If the earth were only one-half as dense as she now is, it would reduce the weight of bodies at her surface one-half. So if a body were taken from the earth's surface half way down to her center, the weight would be reduced one-half. At her center it would be nothing, because the attractive force would be the same in all directions.

In this cut, the diameter of the earth is divided into four equal parts—C, D, E, and F. At A, the whole attraction amounts to four pounds. When the stone reaches B, the part C attracts as strongly upward as D does downward, and their forces balance each other. Then as C and D mutually neutralize each other, we have only the parts E and F, or one-half the globe, to attract the stone downward; consequently the attractive force would be only half as great at B as at A, and the stone would weigh only two pounds.

69. The force with which bodies gravitate toward each other is in direct proportion to their respective masses, and in inverse proportion to the squares of their distances.

A man carried upward in a balloon weighs less and less as his distance from the earth is increased. The same law holds good in regard to the planetary worlds. The nearer a planet is to the sun, or to any other body, the stronger the mutual gravitation.

---

68. How effect weight of bodies on the earth to reduce her density one-half? How to take down half way to center? Quite to center? (Illustrate by diagram.)

69. Give the exact law of gravitation? (What said of a man ascending in a balloon? Of more distant planets?)

70. This great law was discovered by *Sir Isaac Newton*, in 1666. He was then only twenty-four years of age.

<small>The inquiry which led to the discovery is said to have been suggested to the mind of this youthful philosopher by seeing an apple fall from the limb of a tree. "What drew these two globes (the apple and the earth) together?"</small>

PERIODIC REVOLUTIONS OF THE PLANETS.

71. The planets all revolve around the sun from west to east, or toward that part of the heavens in which the sun appears to rise.

<small>To assist his conception of the *direction* in which the planets revolve, the student may suppose that if the earth was in her orbit beyond the sun, at 12 o'clock, she would go what we should call *eastward*, which would be the same direction that we should call westward on the earth, at the same time; as bodies revolving in a circle move in opposite directions on opposite sides of the circle.</small>

72. The passage of a planet from any particular point in its orbit, around to the same point again, is called its *periodic revolution;* and the time occupied in making such revolution is called its *period*, or *periodic time.* The periodic times of the principal planets are as follows:

	Years.	Days.		Years.	Days.
Mercury	0	88	Jupiter	11	317
Venus	0	225	Saturn	29	175
Earth	1	—	Uranus	84	27
Mars	1	322	Neptune	164	226

The Asteroids revolve, on an average, in about $4\tfrac{1}{2}$ years

<small>1. The exact periods of the Asteroids, respectively, are as follows:</small>

	Years.	Days.		Years.	Days.
Flora	3	98	Astræa	4	51
Clio	3	207	Irene	4	55
Vesta	3	230	Eunomia	4	114
Iris	3	250	Juno	4	134
Metis	3	253	Ceres	4	220
Hebe	3	284	Pallas	4	225
Parthenope	3	306	Hygeia	5	217
Egeria	4	45			

<small>2. The periodic time of a planet may very properly be called its *year;* hence a year of Saturn is equal to about thirty of ours, &c. But this difference in the length of the years of the several planets is not owing solely to the difference in the extent of their orbits. There is an actual difference in their velocities, as will be shown in the next paragraph.</small>

<small>70. When and by whom were the Laws of Gravitation discovered? How old? (What led to this discovery?)

71. In what *direction* do the planets revolve in their orbits? (Give illustration.)

72. What meant by the *periodic revolution* of a planet? Its *period* or *periodic time?* Give the periods of the principal planets. Of the Asteroids? (What constitutes the *year* of a planet? Compare years.)</small>

HOURLY MOTION OF THE PLANETS IN THEIR ORBITS.

73. The velocity with which the planets fly through space, in performing their periodical journeys around the sun, varies from 11,000 to 110,000 miles an hour. The hourly motion of the earth amounts to 68,000 miles!

1. The hourly motion of the planets is, approximately, as follows:

	Miles		Miles
Mercury	110,000	Jupiter	30,000
Venus	75,000	Saturn	22,000
Earth	68,000	Uranus	15,000
Mars	55,000	Neptune	11,000

Here, instead of finding the swiftest planets performing the longest periodic journeys, this order is reversed, and they are found revolving in the smallest orbits. The nearer a planet is to the sun, the more rapid its motion, and the shorter its periodic time. The reasons for this difference in the velocities and periodic times of the planets, will appear in a subsequent paragraph.

2. It may seem incredible to the student that the ponderous globe is flying through space at the rate of 68,000 miles an hour, or some 80 times as swift as a bullet; but, like many other astonishing facts in Astronomy, its truth can easily be demonstrated. The diameter of a circle is to its circumference as 7 is to 22 nearly. The earth's distance from the sun being 95,000,000 miles, it is obvious that the whole diameter of her orbit is twice that distance, or 190,000,000; then, as 7 : 22 : : 190,000,000 : 597,142,857 miles, the circumference of the earth's orbit. Divide this sum by 8,766, the number of hours in a year, and we have 68,108 miles as the hourly velocity of the earth.

3. As the earth is not propelled by machinery like a steamboat, or borne upon wheels like a railroad car, it is not strange that we are insensible of its rapid motion, especially as every thing upon its surface, and the atmosphere by which it is surrounded, move onward with it in its rapid flight.

CENTRIPETAL AND CENTRIFUGAL FORCES.

74. The mutual attractive force of the sun and planets is called the *centripetal* force; while the tendency of the planets to fly off from the sun, as they revolve around him, is called the *centrifugal* force.

1. The term *centripetal* is from *centrum*, center, and *peto*, to move toward; and *centrifugal*, from *centrum*, and *fugio*, to fly from the center.

2. The *centrifugal* force is generated by the revolution of the planet, and is in proportion to its velocity—the more rapid the revolution, the stronger the tendency to fly off from the sun.

3. If the centrifugal force were suspended, the planets would at once fall to the sun; and if the centripetal force were destroyed, the planets would fly off in straight lines, and leave the solar system forever. Then might be realized the chaos and confusion of the poet:

> "Let Earth unbalanced from her orbit fly,
> Planets and suns run lawless through the sky;
> Let ruling angels from their spheres be hurled,
> Being on being wreck'd, and world on world."

75. It has already been stated (65), that the force of attraction depends somewhat upon the distances of the attracting bodies—those nearest together being mutually

---

73. What said of the *velocity* of the planets? Of the earth? (Table? Remarks upon it? How is the hourly velocity of the earth ascertained? Why not sensible of this rapid motion?)

74. Centripetal and centrifugal forces? (Derivation of terms? How centrifugal force generated? Suppose either suspended?)

attracted most. It follows, therefore, that Mercury has the strongest tendency toward the sun, Venus next, the Earth next, &c., till we get through to Neptune; and as the centrifugal force which is to balance the centripetal is created by the velocity or projectile force of the planets, that velocity must needs be in proportion to their distances, respectively, from the sun; the nearest revolving the most rapidly. This we find to be the actual state of things in the solar system.

The mechanism of the solar system strikingly displays the wisdom of the great Creator. The centrifugal force depends, of course, upon the rapidity of the revolution; and in order that these forces might be exactly balanced, God has imparted to each planet a velocity just sufficient to produce a centrifugal force equal to that of its gravitation. Thus they neither fall to the sun on the one hand, nor fly off beyond the reach of his beams on the other, but remain balanced in their orbits between these two great forces, and steadily revolving from age to age. "How manifold are thy works! in wisdom hast thou made them all."

LAWS OF PLANETARY MOTION.

76. Three very important laws, or principles, governing the movements of the planets, were discovered by *Kepler*, a German astronomer, in 1609. In honor of their discoverer, they are called *Kepler's Laws.*

Kepler was a disciple of Tycho Brahe, a noted astronomer of Denmark, and was equally celebrated with his renowned tutor. His residence and observatory were at Wirtemburgh, in Germany.

77. The *first* of these laws is, that *the orbits of all the planets are elliptical, having the sun in the common focus.*

The point in a planet's orbit nearest the sun is called the *perihelion* point, and the point most remote the *aphelion* point. *Perihelion* is from *peri*, about or near, and *helios*, the sun; and *aphelion*, from *apo*, from, and *helios*, the sun.

From this first law of Kepler, it results that the planets move with different velocities, in different parts of their orbits. From the aphelion to the perihelion points, the centripetal force *combines with* the centrifugal to accelerate the planet's motion; while from perihelion to

---

75. Why the planets nearest the sun revolve most rapidly in their orbits. (Remark?)
76. Laws of planetary motion? (Who was *Kepler*?)
77. State the *first* of Kepler's laws. Perihelion? Aphelion?

aphelion points, the centripetal acts *against* the centrifugal force, and *retards* it.

1. From A to B in the diagram, the centrifugal force, represented by the line C, acts *with* the tendency to revolve, and the planet's motion is *accelerated*; but from B to A the same force, shown by the line D, acts *against* the tendency to advance, and the planet is *retarded*. Hence it comes to Aphelion with its least velocity, and to Perihelion with its greatest.

2. In the statement of velocities on page 45, the *mean* or *average* velocity is given.

78. The *second* law is, that the *radius vector of a planet describes equal areas in equal times*. The radius is an imaginary line joining the center of the sun and the center of the planet, in any part of its orbit. *Vector* is from *veho*, to carry; hence the *radius vector* is a radius carried round. By the statement that *it describes equal areas in equal times*, is meant that it sweeps over the same surface in an hour, when a planet is near the sun, and moves swiftly, as, when furthest from the sun, it moves most slowly.

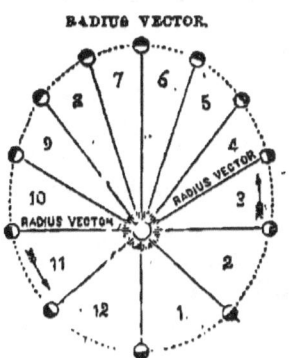

RADIUS VECTOR.

The nearer a planet is to the sun, the more rapid its motion. It follows, therefore, that if the orbit of a planet is an ellipse, with the sun in one of the foci, its rate of motion will be unequal in different parts of its orbit—swiftest at perihelion, and slowest at aphelion. From perihelion to aphelion the centripetal more directly counteracts the centrifugal force, and the planet is *retarded*. On the other hand, from the aphelion to the perihelion point, the centripetal and centrifugal forces are united, or act in a similar direction. They consequently hasten the planet onward, and its rate of motion is constantly *accelerated*. Now suppose, when the planet is at a certain point near its perihelion, we draw a line from its center to the center of the sun. This line is the *radius vector*. At the end of one day, for instance, after the planet has advanced considerably in its orbit, we draw another line in the same manner to the sun's center, and estimate the area between the two lines. At another time, when the planet is near its aphelion, we note the space over which the radius vector travels in one day, and estimate its area. On comparison, it will be found, that notwithstanding the unequal *velocity* of the planet, and consequently of the radius vector, at the two ends of the ellipse, the *area* over which the radius vector has traveled is the same in both cases. The same principle obtains in every part of the planetary orbits, whatever may be their ellipticity or the mean distance of the planet from the sun; hence the rule, that *the radius vector describes equal areas in equal times*. In the preceding cut, the twelve triangles, numbered 1, 2, 2, &c., over each of which the radius vector sweeps in equal times, are equal.

79. The *third* law of Kepler is, that *the squares of the*

---

78. State the *second* law of planetary motion. Define *radius vector*. (Explain this second law.)

periodic times of any two planets are proportioned to the cubes of their mean distances from the sun.

1. Take, for example, the earth and Mars, whose periods are 365·2564 and 686·9796 days, and whose distances from the sun are in the proportion of 1 to 1·52369, and it will be found that $(365·2564)^2 : (686·9796)^2 :: (1)^3 : (1·52369)^3$.
2. According to these laws, which are known to prevail throughout the solar system, many of the facts of astronomy are deduced from other facts previously ascertained. They are, therefore, of great importance, and should be studied till they are, at least, thoroughly understood, if not committed to memory.

ASPECTS OF THE PLANETS.

80. By the *aspects* of the planets is meant their positions in their orbits with respect to each other. The principal aspects are *conjunction, quadrature,* and *opposition.* Two bodies are in *conjunction* when in the same longitude; that is, on the same north and south line in the heavens. The sign for conjunction is ☌. When 90° apart, bodies are said to be in *quadrature,* with the sign □; and when 180° apart, or in opposite parts of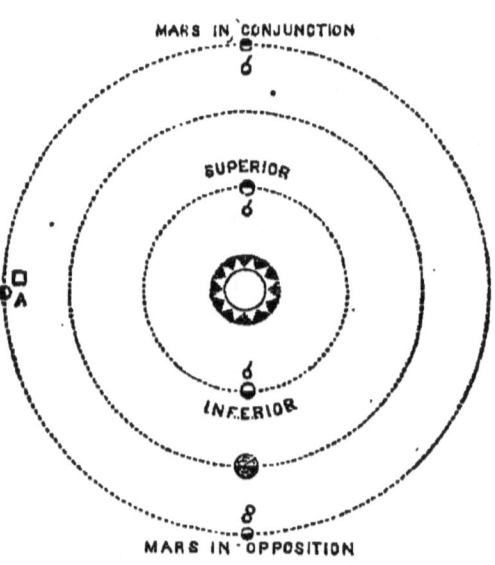
the heavens, they are in opposition, and the sign is ☍. Conjunctions are of two kinds. An *inferior conjunction* is when the planet is between the earth and the sun; and a *superior conjunction,* when it is beyond the sun.

1. Let the student imagine himself stationed upon the earth in the cut. Then the sun and three planets above are in *conjunction.* The *inferior* and *superior* are distinguished; while at A, a planet is shown in *quadrature,* and at the bottom of the cut the planet Mars in *opposition* with the sun and interior planet.

79. State the *third* law. (Illustration? What said of the importance of a knowledge of these laws?)
80. What meant by the *aspects* of the planets? State principal aspects? Define each. Signs? How many kinds of conjunctions? Define each. (Explain by diagram. When is Venus nearest? What difference at superior and inferior conjunctions?)

2. When at her superior conjunction, Venus is 164 millions of miles from the earth; but when at her inferior conjunction, she is only 26 millions of miles distant, or the whole diameter of her orbit nearer.

SIDEREAL AND SYNODIC REVOLUTIONS.

81. The *sidereal revolution* of a planet is a complete revolution from any given point in its orbit around to the same point again.

Sidereal, from *sideralis*—a revolution as measured by the stars. See page 28, note 1. The periodic revolutions of the planets, given at Art. 71, are sidereal revolutions.

82. A *synodic revolution* is from one conjunction to the same conjunction again.

1. The term *synod* signifies a *meeting* or *convention;* and the synodic revolution of a planet is a *meeting* revolution: that is, from one meeting or conjunction to another.
2. The difference between a sidereal and synodic revolution may be illustrated by the motion of the hands of a clock or watch. At twelve o'clock, the hour and minute hands are together; but at one o'clock, when the minute-hand has made a complete revolution, and points to XII. again, the hour-hand has gone forward to I., and the minute-hand will not overtake it till about four minutes afterward. The revolution of the minute-hand from XII. to XII. again, represents the *sidereal* revolution of a planet; and when it overtakes the hour-hand, it becomes a *synodic* revolution.
3. The sidereal and synodic periods of the principal planets are as follows:

	Sidereal.			Synodic.
Mercury	—	88 days		115 days
Venus	—	225 "		594 "
Mars	1 year,	322 "		780 "
Jupiter	11 "	317 "		399 "
Saturn	29 "	175 "		378 "
Uranus	84 "	—		367½ "
Neptune	164 "	226 "		367½ "

From this table it is seen that the synodic periods of the more distant planets correspond very nearly with the periodic time of the earth. Being remote from the sun, they move very slowly, and the earth soon overtakes them, after performing her periodic revolution.

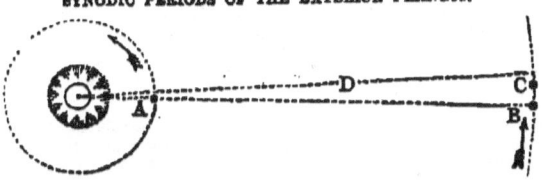

SYNODIC PERIODS OF THE EXTERIOR PLANETS.

Suppose the earth and Uranus to be in conjunction, as shown at A B. In 365¼ days, the earth performs her sidereal or periodic revolution, and returns to the point A again. In the mean time Uranus, whose periodic time is 84 years, has passed

---

81. What meant by the *sidereal* revolution of a planet? (Derivation of term? Are the periods of the planets sidereal revolutions?)
82. *Synodic* revolution? (Illustrate difference by clock. What fact respecting synodic periods of distant planets? How explained? Illustrate by diagram.)

through only $\frac{1}{84}$th part of his orbit, or about $4\frac{1}{4}°$ to the point C; and in $4\frac{1}{4}$ days the earth overtakes him on the line D. It is on this account that the synodic period of Uranus is only $367\frac{1}{4}$ days, or $4\frac{1}{4}$ days longer than the periodic time of the earth.

THE ECLIPTIC, ZODIAC, SIGNS, ETC.

83. The *Ecliptic* is *the plane of the earth's orbit*, or the path in which the sun appears to revolve in the heavens.

PLANE OF THE ECLIPTIC.

1. In the above cut, an attempt is made to represent the ecliptic, or plane of the earth's orbit. It is an oblique view, which makes the orbit appear elliptical. It shows one-half of the sun and half the earth on one side, and half on the other. The circle projecting beyond the orbit is to represent the plane of the ecliptic, indefinitely extended.

2. If the student has any difficulty in getting a correct idea respecting the ecliptic, let him suppose the orbit of the earth to be a hoop of small wire laid upon a table: the surface of the table, both within and without the hoop, would then represent the plane of the ecliptic. From the above definition and description, it will be seen that the ecliptic passes through the center of the earth, and the center of the sun; consequently the ecliptic and the apparent path of the sun through the heavens are in the same plane. It will be easy, therefore, to ascertain the true position of the ecliptic in the heavens, and to imagine its course among the stars.

3. The plane of the earth's orbit is called the *ecliptic*, because *eclipses* of the sun and moon never take place except when the moon is in or near this plane.

84. The position of the ecliptic to persons north of the equator is *south* of us. It runs east and west, cutting the centers of the sun and earth. *North* of the ecliptic is called *above it;* and *south* of it, *below it.*

The student should again be reminded that there is no absolute *up* or *down* in the universe. He must also guard against the idea that the ecliptic may be *horizontal*. This term has reference only to the earth, and is descriptive of a place depending altogether for its own position upon that of the observer, as shown and illustrated at 20. Though the ecliptic is a permanent plane, and cuts the starry heavens around us at the same points from age to age, it has no absolute *up* or *down*, unless it should be the direction to and from the sun. The distinction of *above* and *below* is merely arbitrary, and grows out of our position north of the equator, which makes the south side of the ecliptic appear down to us.

---

83. What the *ecliptic?* (How cut the earth and sun? Point out its course in the heavens. Why called the *ecliptic?*)

84. What meant by *above* and *below* the ecliptic? (Remarks in note.)

ECLIPTIC, ZODIAC, SIGNS, ETC.   51

85. The *Poles of the Ecliptic* are the extremities of an imaginary axis upon which the ecliptic seems to revolve.

As the ecliptic and equinoctial are not in the same plane, their poles do not coincide, or are not in the same points in the heavens. The cause of this variation will be explained hereafter.

86. The *Zodiac* is an imaginary belt 16° wide, viz., 8° on each side of the ecliptic, and extending from west to east quite around the heavens. In the heavens, it includes the sun's apparent path, and a space of eight degrees south, and eight degrees north of it.

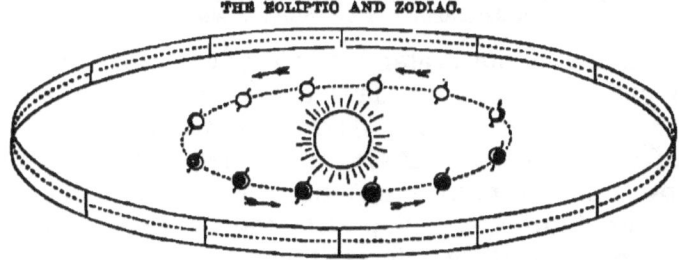

THE ECLIPTIC AND ZODIAC.

In this cut, the interior dotted circle represents the earth's orbit; the exterior the *plane* of her orbit extended to the starry heavens. The dark lines each side of the ecliptic are the limits of the zodiac. The earth is shown in perspective, largest near to us, and growing smaller as her distance is increased. The *arrows* show her direction.

87. The great circle of the zodiac is divided into twelve equal parts, called *signs*. (These divisions are shown in the above cut, by the spaces between the perpendicular lines that cross the zodiac.) The ancients imagined the stars of each sign to represent some animal or object, and gave them names accordingly. On this account, they gave the name *zodiac* to this belt around the heavens; not, as some have imagined, because it was a *zone*, but from the Greek *zoön*, an animal, because so many animals were represented within its limits.

---

85. The *poles* of the ecliptic? (Do the poles of the ecliptic and the poles of heavens coincide?)
86. What is the *zodiac?*
87. How is the zodiac divided? Idea of the ancients? Origin of the name *zodiac?*

88. The names, order, and symbols of the twelve signs of the zodiac are as follows:

Aries (or the Ram)....	♈	Libra (the Balance)....	♎
Taurus (the Bull).....	♉	Scorpio (the Scorpion).	♏
Gemini (the Twins)....	♊	Sagittarius (the Archer)	♐
Cancer (the Crab).....	♋	Capricornus (the Goat)	♑
Leo (the Lion)........	♌	Aquarius (the Waterman)	♒
Virgo (the Virgin)....	♍	Pisces (the Fishes)....	♓

These names being from the Latin, their signification is added in parentheses, and should be understood by the pupil. In reciting, however, it is only necessary to give the *first names*—as Aries, Taurus, Gemini, &c. By carefully observing these symbols, the student will detect a resemblance between several of them and the objects they represent. For instance, the sign for Aries represents his *horns;* so also with Taurus, &c.

89. The ancients pretended to predict future events by the signs, aspects, &c. This art, as it was called, was denominated *Astrology.* Astrology was either *natural* or *judicial.* *Natural Astrology* aimed at predicting remarkable occurrences in the natural world, as earthquakes, volcanoes, tempests, and pestilential diseases. *Judicial Astrology* aimed at foretelling the fates of individuals or of empires.

ANCIENT ASTROLOGY.

"This science," says Webster, "was formerly in great request, as men ignorantly supposed the heavenly bodies to have a ruling influence over the physical and moral world; but it is now universally exploded by true science and philosophy." A fragment of this ancient superstition, like the adjoining figure, may still be met with occasionally in the pages of an almanac; and there are still persons to be found in almost every community who think certain "signs" govern certain portions of the human body, and that it is very important to do every thing "when the sign is right." Impostors, also, are still taking advantage of this credulity; and, professing to "tell fortunes," as they call it, by the stars, impose upon and defraud the ignorant. The stars have no more to do with our "destiny" than we have with theirs.

90. The *order* of the signs is from west to east around the heavens. Thus Aries, Taurus, Gemini, &c., around to Pisces.

---

88. *Names* of the signs? Symbols on blackboard.
89. What is *astrology?* How divided? Define each. (Remark of Webster? Of the author?)
90. The *order* of the signs? (Describe the cut. What said of Taurus?)

CELESTIAL LATITUDE AND LONGITUDE. 53

PERPENDICULAR VIEW OF THE ECLIPTIC.

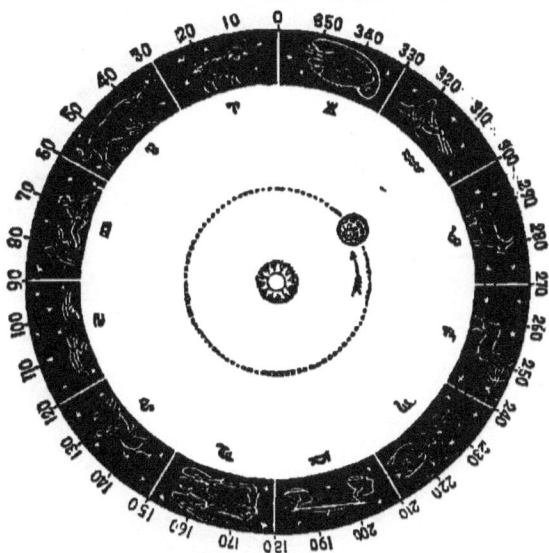

On pages 50 and 51, we presented *oblique* views of the ecliptic. The above is a *perpendicular* view. The sun is seen in the center, and the earth revolving around him; and in the distance is shown the circle of the starry heavens. This circle is divided into twelve equal parts, representing the twelve signs; while the *object* which the stars in each sign were supposed to resemble is placed in that sign, and the *symbol* immediately opposite and within the sign. But the head of Taurus should point east instead of west.

CELESTIAL LATITUDE AND LONGITUDE.

91. *Celestial Longitude* is distance *east* of a given point in the heavens, reckoned on the *ecliptic*. Beginning at the *Vernal Equinox*, it is reckoned eastward to 360°, or to the point whence we started.

<small>The pupil will consult the preceding cut, in which the longitude is marked for every ten degrees. By holding the book up to the south of him, the surface of the page will represent the plane of the ecliptic; and the reckoning of 10, 20, 30, &c., from the top of the cut *eastward*, will answer to the manner in which celestial longitude is reckoned eastward around the heavens.</small>

92. Celestial Longitude is either *Heliocentric* or *Geocentric*. The *heliocentric* longitude of a planet is its longitude as viewed from the sun; and the *geocentric*, its longitude as viewed from the earth.

<small>*Geocentric* is from *ge*, the earth, and *kentron*, center; and *heliocentric* from *helios*, the sun, and *kentron*, center.</small>

---

91. Celestial longitude? Where begin to reckon? Illustrate by book. Point out order of reckoning in the heavens.
92. What is *heliocentric* longitude? *Geocentric*? (Derivation of terms? Illustrate by diagram.)

GEOCENTRIC AND HELIOCENTRIC LONGITUDE.

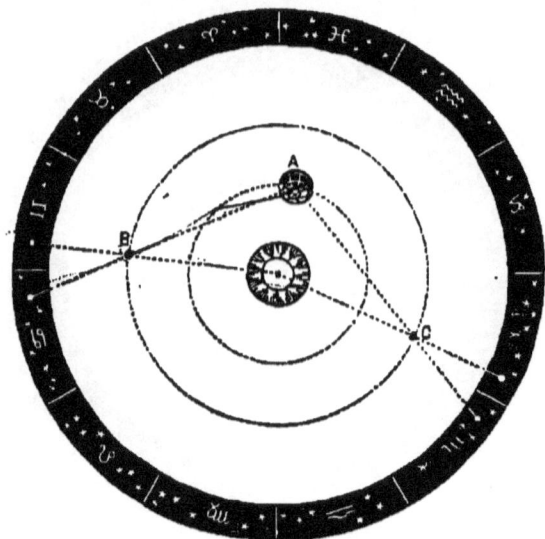

In this cut, the planet B, when viewed from the earth at A, seems to be in the sign ♋; but when viewed from the sun, it appears to be in ♊. Again: when at C, her apparent longitude from the earth is in ♏; when from the sun, she appears to be in ♐. The learner will not only perceive the difference between *geocentric* and *heliocentric* longitude, but will see why the latter more than the former indicates the true position of the planet. It is an easy thing, however, if one is known, to deduce the other from it.

MEAN AND TRUE PLACES OF A PLANET.

93. The *mean place* of a planet is the place it would have occupied had it revolved in a circular orbit, and with uniform velocity.

The *true place* is that which it really occupies, revolving as it does in an elliptical orbit, and with unequal velocity.

1. In the cut, the dotted ellipse represents the orbit of the planet, and the points T T T, &c., its *true place*. In the circle or hypothetical orbit, the points M M, &c., indicate the *mean place* of the planet.

MEAN AND TRUE PLACES OF A PLANET

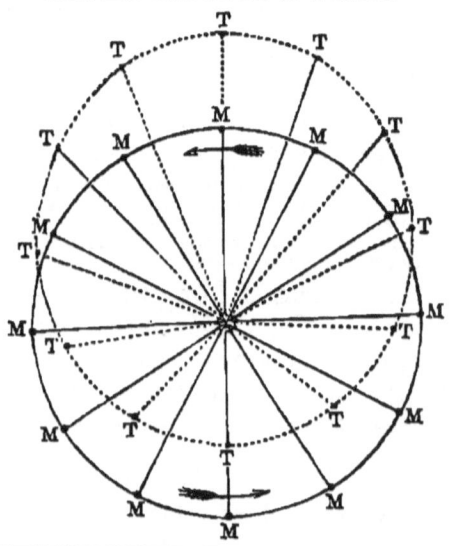

93. What is meant by the *mean place* of a planet? The *true place*? (When

DIRECT AND RETROGRADE MOTIONS. 55

2. From the *perihelion* to the *aphelion*, it will be seen that the true place is in advance, or *eastward*, of the mean place; while from *aphelion* to *perihelion* again, the mean place is in advance of the true. But at the perihelion and aphelion points, the mean and true places coincide.

3. In one respect, the cut conveys an erroneous impression, as it represents the planet as passing over an equal distance in its orbit in equal times. This is not the fact. The difference in its velocity in different parts of its orbit could not well be represented here; but the student will find it beautifully illustrated by the second cut on page 47, and in the explanatory note accompanying it.

94. *Celestial Latitude* is distance *north* or *south* of the *ecliptic*, and is reckoned to the pole of the ecliptic, or to 90°.

DIRECT AND RETROGRADE MOTIONS.

95. The apparent motion of a planet is said to be *direct* when it is *eastward* among the stars, and *retrograde* when it seems to go back or *westward* in the ecliptic. When it seems to move neither east nor west, it is said to be *stationary*.

96. The cause of the apparent retrogression of the *interior* planets is the fact that they revolve much more rapidly than the earth, from which we view them; causing their direct motion to appear to be retrograde.

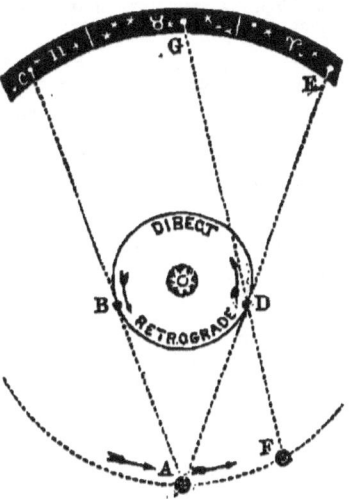

DIRECT AND RETROGRADE MOTIONS.

Suppose the earth to be at A, and Venus at B, she would appear to be at C, among the stars. If the earth remained at A while Venus was passing from B to D, she would seem to retrograde from C to E; but as the earth passes from A to F while Venus goes from B to D, Venus will appear to be at G; and the amount of her apparent westward motion will only be from C to G.

97. The apparent retrograde motions of the *exterior* planets is due to the rapidity with which the position from which we view them is

---

is the true in advance of the mean? When the reverse? When do they coincide? Wherein is the cut defective? Where have we a true representation?)

94. Celestial *latitude*? How reckoned?

95. When is a planet's apparent motion *direct? Retrograde?* When is a planet said to be *stationary?*

96. State the *cause* of the apparent retrogression of an *interior* planet. (Illustrate by diagram.)

97. The cause of retrogression of *exterior* planets. (Illustrate by diagram. What fact shown?)

changed, as we are carried rapidly through space with the earth, in her annual journey around the sun.

98. The portion of the ecliptic through which a planet seems to retrograde is called the *Arc of Retrogradation.* The more remote the planet the less the arc, and the longer the time of its retrogression.

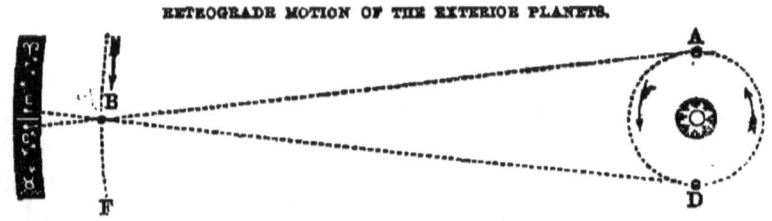

RETROGRADE MOTION OF THE EXTERIOR PLANETS.

1. Suppose the earth at A, and the planet Neptune at B, he would then appear to be at C, among the stars; but as Neptune moves but a little from B toward F, while the earth is passing from A to D, Neptune will appear to retrograde from C to E. Whatever Neptune may have moved, however, from B toward F, will go to reduce the amount of apparent retrogression.

2. It is obvious from this figure, that the more distant an exterior planet is, and the slower it moves, the less will be its arc of retrogradation, and the longer will it be retrograding. Neptune appears to retrograde 180 days, or nearly half the year.

The following table exhibits the *amount of arc* and the *time* of the retrogradation of the principal planets:

	Arc.	Days
Mercury	13½°	23
Venus	16	42
Mars	16	73
Jupiter	10	121
Saturn	6	139
Uranus	4	151
Neptune	1	180

99. The *greatest elongation* of an interior planet is the greatest apparent distance east or west of the sun at which it is ever found.

In the second cut back, the point B would represent the greatest *eastern*, and D the greatest *western*, elongation of the planet. At these two points she would appear to be stationary.

100. The greatest elongations of Venus vary from 45° to 48°. The fact that she never departs more than 48° from the sun proves that her orbit is *within* that of the earth; and the variation in her elongations shows that her orbit is not an exact circle.

---

98. What meant by the *Arc of Retrogradation?*
99. Greatest elongation?
100. Greatest elongation of Venus? What does it prove?

101. When Venus is *west* of the sun, and rises before him, she is *morning star;* and when *east* of the sun she is *evening star.*

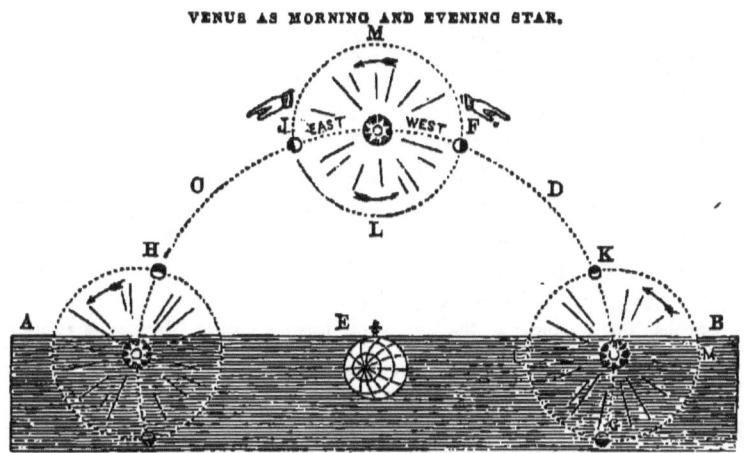

VENUS AS MORNING AND EVENING STAR.

1. Let the student hold the book up *south* of him, and he will at once see why Venus is alternately morning and evening star. Let the plane A B represent the sensible or visible horizon, C D the apparent daily path of the sun through the heavens, and E the earth in her apparent position. The sun is shown at three different points— namely, rising in the east, on the meridian, and setting in the west; while Venus is seen revolving around him from west to east, or in the direction of the arrows. Now it is obvious that when Venus is at F, or *west* of the sun, she sets before him as at G, and rises before him as at H. She must, therefore, be *morning star.* On the other hand, when she is *east* of the sun, as at J, she lingers in the west after the sun has gone down, as at K, and is consequently *evening star.*

2. In this cut, Venus would be at her greatest elongation *eastward* at J, and *westward* at F, and in both cases would be "*stationary.*" At L and M she would be in *conjunction* with the sun.

3. Were the earth to suspend her daily rotation, with the sun on the meridian of the observer, as represented at L, we might readily watch Venus through her whole circuit around the sun.

4. Venus may sometimes be seen at mid-day, either east or west of the sun, and Dr. Dick considers the day-time most favorable for observing her with a telescope.

102. Venus is morning and evening star, alternately, for about 292 days, or from one conjunction to another. Appearing first east and then west of the sun, she was regarded by the ancients as *two different stars,* which they called *Phosphor* and *Hesperus.*

When Venus is near her greatest elongation from the sun, she is one of the most beautiful stars in the heavens. She is very easily found, either just before sunrise, or just after sundown; and we earnestly recommend the class to ascertain where she is, at the time of learning this lesson, and to watch her movements for a few months, and see if they do not correspond with the description here given. The knowledge acquired will thus be *located* in the heavens.

---

101. When *morning* and when *evening* star?
102. How long is Venus alternately morning and evening star? How regarded by the ancients? (Remark in note?)

103. The greatest elongations of Mercury vary from 16 to 29 degrees from the sun, which proves his orbit to be elliptical, and to be within that of Venus.

1. As Mercury *never* departs more than 29° from the sun, when at his greatest elongation, and Venus is never *nearer* than about 45°, when at her greatest elongation, it is evident that his orbit is inside that of Venus.
2. When at perihelion, Mercury is only 29,305,000 miles from the sun's center; while in the opposite part of his orbit, or in aphelion, he reaches to 44,474,000—making a variation of distance, arising from the ellipticity of his orbit, of more than 15,169,000 miles, which is nearly five times as great as in the case of the earth.

104. In consequence of the nearness of Mercury to the sun, he is very rarely seen; and if seen at all, it must be in strong twilight, either morning or evening. He never appears conspicuous, even under the most favorable circumstances, but twinkles like a star of the third magnitude, with a pale rosy light.

By consulting an almanac, the student can ascertain when Mercury is at his greatest elongation, and if it is *eastward*, look out for him low down in the west, just after sunset. If his elongation is *westward*, he must be looked for *in the east*, before sunrise. It will be worth rising early to see him.

DEVIATION OF THE ORBITS OF THE PLANETS FROM THE PLANE OF THE ECLIPTIC.

105. Although the sun is the great center around which all the planets revolve, it should be borne in mind that *no two of them revolve in the same plane*. Taking the plane of the earth's orbit or ecliptic as the standard, the orbits of the other planets all depart from that plane, some more and some less. As a consequence, they all pass through or cut the plane of the ecliptic twice at every revolution.

VENUS PASSING AND REPASSING THE PLANE OF THE EARTH'S ORBIT.

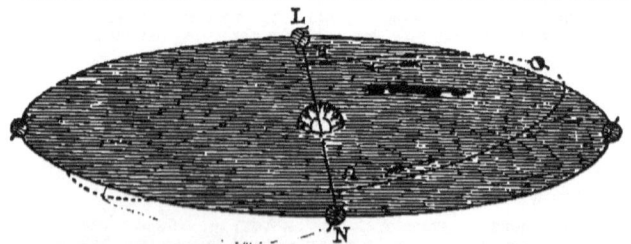

In this cut, the space included within the orbit of the earth is united to represent a *plane*. Within her orbit, and part above, and part below it, may be seen the orbit of

---

103. Greatest elongation of *Mercury?* Proves what? (Show how demonstrated. What said of the eccentricity of the orbit of Mercury?)
104. Is Mercury often seen? When, if at all? Appearance? (How find?)
105. Are all the planetary orbits in the same plane? What consequence follows?

DEVIATION OF THE ORBITS OF THE PLANETS.

Venus, the arrows showing her direction. Her orbit goes out of sight when it passes the plane of the ecliptic.

106. The points where a planet passes the plane of the ecliptic are called the *Nodes* of its orbit. They are in opposite sides of the ecliptic, and of course 180° apart. The point where they pass *south* of the ecliptic is called the *descending* node, and marked ☋ ; and that through which they pass *north* of the ecliptic is called the *ascending* node, and marked ☊. The *Line of the Nodes* is a line drawn from one node to the other across the ecliptic. The nodes, ascending and descending, and their symbols, and also the line of the nodes, marked L N, are all well represented in the preceding cut.

INCLINATION OF THE ORBITS OF THE PLANETS TO THE PLANE OF THE ECLIPTIC.

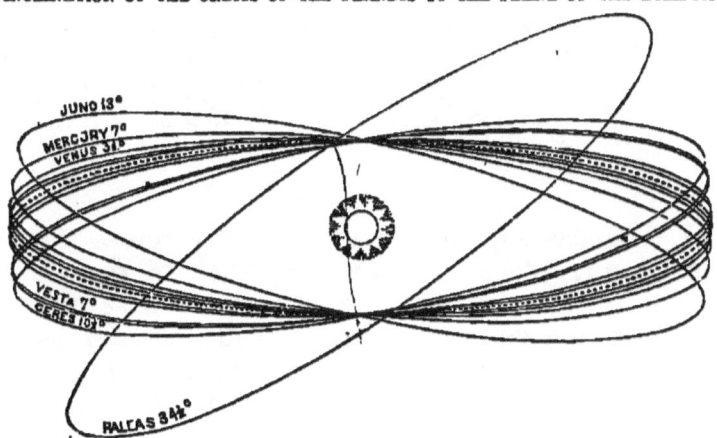

107. The nodes of the planetary orbits are not all in the same longitude, but are distributed all around the ecliptic. In astronomical works and calculations, the longitude of the ascending node only is noted, as the opposite node is always just 180° from it.

The longitude of the ascending nodes of the planets, respectively, is as follows:

Mercury	46°	Metis	69°	Ceres	81°
Venus	75	Hebe	138	Pallas	173
Earth	—	Parthenope	—	Hygeia	—
Mars	48	Egeria	—	Jupiter	98
Flora	110	Astræa	141	Saturn	112
Clio	—	Irene	—	Uranus	72
Vesta	108	Eunomia	—	Neptune	130
Iris	260	Juno	171		

106. What are the *Nodes* of a planet's orbit? How situated with respect to each other? What called respectively, and why? What meant by the *line* of the nodes?

107. Are the nodes of all the planetary orbits in the same longitude? How distributed? Which node usually mentioned and located? Why not both?

108. The deviation of the planets, respectively, from the ecliptic varies from 1° 46″ to 34½°. The orbits of the larger planets are all near the ecliptic, while some of the asteroids depart widely from it. On this account they are sometimes called *ultra-zodiacal* planets.

<small>The preceding cut may help the student to form an idea of the inclination of the planetary orbits; but we must guard against the impression it may make that all the planetary *nodes are in the same part of the ecliptic*, as we were obliged to represent in the cut. Instead of this, they are distributed all about the ecliptic. Again: the cut shows the several planets at about the same *distance* from the sun, contrary to the fact, as stated and illustrated on page 80. The *dotted* line represents the earth's orbit, or plane of the ecliptic, and the other lines the planes of the orbits of several of the planets, and their departure from the ecliptic. The inclination of the several orbits is, in round numbers, as follows:</small>

Mercury	7°	Metis	5° 34′	Ceres	10° 37′
Venus	3° 23′	Hebe	14° 47′	Pallas	34° 37′
Earth	—	Parthenope	—	Hygeia	—
Mars	1° 51′	Egeria	—	Jupiter	1° 18′
Flora	5° 53′	Astræa	5° 19′	Saturn	2° 29′
Clio	7° 08′	Irene	—	Uranus	1° 46′
Vesta	7° 08′	Ennomia	—	Neptune	1° 46′
Iris	5° 28′	Juno	18° 3′		

OF TRANSITS.

109. The passage of a heavenly body across the meridian of any place, or across the disk of the sun, is called a *transit*. A planet will seem to pass over the disk of the sun when it passes directly *between* us and him; and as none but the interior planets can ever get between us and the sun, it is obvious that no others can ever make a transit over his disk.

<small>The term *transit* is sometimes used with reference to terrestrial objects, as when we speak of the *transit* or passage of goods through a country. The words *transition, transitive, transitory*, &c., are derived from the primitive word *transit*.</small>

110. Mercury and Venus are the only planets that can appear to cross over the sun's disk, as viewed from our globe.

<small>Were we stationed upon one of the remote exterior planets, we might see the earth, and Mars, and Jupiter transit the sun; but as it is, we shall never witness such phenomena, or, at least, till we leave the present world.</small>

111. Were the orbits of Mercury and Venus in the same plane with that of the earth, they would transit the

---

<small>108. To what extent do the planetary orbits depart from the ecliptic? What said of the larger planets? Of the smaller? (Remarks upon the cut. State the inclination of Mercury, Venus, Mars, &c.)
109. What is a *transit*? When do planets transit the sun? What planets do this? Why not the exterior? (Remarks upon term transit.)
110. What planets make transits across the sun's disk? (**Remarks in note.**)</small>

sun at every synodic revolution; but as one-half of each of their orbits is *above*, and the other half *below* the ecliptic, they generally appear to pass either above or below the sun.

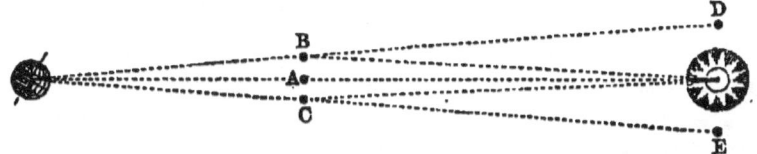

Let the right line A, joining the earth and the sun in the above diagram, represent the plane of the ecliptic. Now when an interior planet is in this plane, as shown at A, it may appear to be upon the sun's disk; but if it is either above or below the ecliptic, as shown at B and C, it will appear to pass either above or below the sun, as shown at D and E.

112. A transit can never occur except when the interior planet is in or very near the ecliptic. The earth and the planet must be on the same side of the ecliptic; the planet being at one of its nodes, and the earth on the line of its nodes.

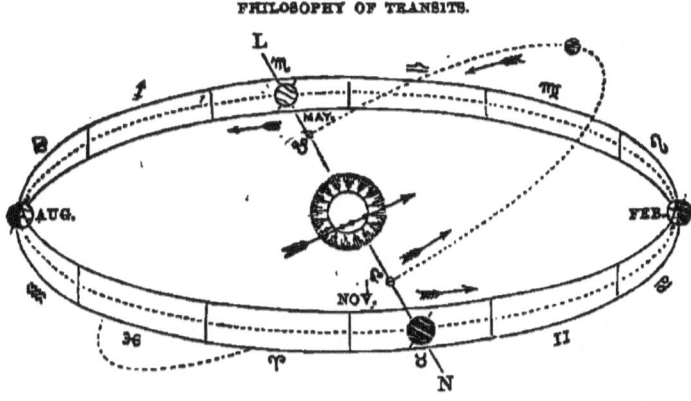

PHILOSOPHY OF TRANSITS.

This cut represents the ecliptic and zodiac, with the orbit of an interior planet, his nodes, &c. The line of his nodes is, as shown, in the 16° of ♉ and the 16° of ♏. Now if the earth is in ♉, on the line L N, as shown in the cut, when Mercury is at his *ascending node* (☊), he will seem to pass *upward* over the sun's face, like a dark spot, as represented in the figure. On the other hand, if Mercury is at his ☋ when the earth is in the 16° of ♏, the former will seem to pass *downward* across the disk of the sun.

113. As the nodes of the planetary orbits are in oppo-

111. Why not transits every revolution of Mercury and Venus? (Illustrate by diagram.)
112. When must transits occur, if at all? Where must the earth and planet be? (Illustrate by diagram.)

site sides of the ecliptic, it follows that the earth must pass the line of the nodes of the interior planets, respectively, in opposite months of the year. These months are called the *node months* of the planet, and are the months in which all its transits must occur.

114. In making transits across the sun's disk, the planets seem to pass from east to west, and to ascend or descend, as respects the ecliptic, according as the planet is at the ascending or descending node.

This variation in the direction of the planets, during different transits, is well represented in the next cut.

115. The node months of Mercury are May and November.

All the transits of Mercury ever noticed have occurred in one or the other of these months, and for the reason already assigned. The first ever observed took place November 6, 1631; since which time there have been 29 others by the same planet—in all 30—8 in May, and 22 in November.

116. The last transit of Mercury occurred November 9, 1848; and the next will take place November 11, 1861. Besides this, there will be five more during the present century—two in May, and three in November.

The accompanying cut is a delineation of all the transits of Mercury from 1802 to the close of the present century. The dark line running east and west across the sun's center represents the plane of the ecliptic, and the dotted lines the apparent paths of Mercury in the several transits. The planet is shown at its nearest point to the sun's center. His path in the last transit and in the next will easily be found.

2. The last transit of Mercury was observed in this country by Professor Mitchel, at the Cincinnati Observatory, and by many others both in America and in Europe.

---

113. What are the *node months*? (Explain by diagram.)
114. In what direction do planets cross the sun in transits, and why?
115. Which are the node months of Mercury?
116. When did the last transit of Mercury occur? When will the next take place? (What represented in the cut? Describe. Where is the planet shown? What said of last transit of Mercury?)

The writer had made all necessary preparation for observing the phenomenon at his residence, near Oswego, New York; but, unfortunately, his sky was overhung with *clouds*, which hid the sun from his view, and disappointed all his hopes.

117. The node months of Venus are December and June. The line of her nodes lies in Gemini (Ⅱ) and Sagittarius (♐); and as the earth always passes those points in the months named, it follows that all transits of Venus must occur in those months for ages to come.

This proposition will be well understood by consulting the cut on page 61; for as the ine of Venus's nodes is only one sign ahead of that of Mercury, the earth will reach hat point in the ecliptic in one month after she passes the line of Mercury's nodes; so hat if his transits occur in May and November, hers should occur in June and December, as is always the case.

118. The last transit of Venus occurred June 3, 1769; and the next will take place December 8, 1874.

1. Only three transits of Venus have as yet been observed—namely, December 4, 1639; June 5, 1761; and June 3, 1769. It is said that Rittenhouse was so interested in viewing that of 1769, that he actually fainted. In defining the term *transit*, Dr. Webster says: "I witnessed the *transit* of Venus over the sun's disk, June 3, 1769." (See "Unabridged" Dictionary.) The next four will occur December 8, 1874; December 5, 1882; June 7, 2004; and June 5, 2012.

2. The first transit ever witnessed was that of December 4, 1639. The observer was a young man named Horrox, living in an obscure village near Liverpool, England. The table of Kepler, constructed upon the observations of Tycho Brahe, indicated a transit of Venus in 1631, but none was observed. Horrox, without much assistance from books and instruments, set himself to inquire into the error of the tables, and found that such a phenomenon might be expected to happen in 1639. He repeated his calculations during this interval with all the carefulness and enthusiasm of a scholar ambitious of being the first to predict and observe a celestial phenomenon which, from the creation of the world, had never been witnessed. Confident of the result, he communicated his expected triumph to a confidential friend residing in Manchester, and desired him to watch for the event, and to take observations. So anxious was Horrox not to fail of witnessing it himself, that he commenced his observations the day before it was expected, and resumed them at the rising of the sun on the morrow. But the *very hour* when his calculations led him to expect the visible appearance of Venus on the sun's disk, *was also the appointed hour for the public worship of God on the Sabbath*. The delay of a few minutes might deprive him forever of an opportunity of observing the transit. If its very commencement were not noticed, clouds might intervene, and conceal it until the sun should set; and nearly a century and a half would elapse before another opportunity would occur. He had been waiting for the event with the most ardent anticipation for eight years, and the result promised much benefit to the science. *Notwithstanding all this, Horrox twice suspended his observations, and twice repaired to the house of God*, the great Author of the bright works he delighted to contemplate. When his duty was thus performed, and he had returned to his chamber the second time, his love of science was gratified with full success, and he saw what no mortal eye had observed before. If any thing can add interest to this incident, it is the modesty with which the young astronomer apologizes to the world for *suspending* his observations at all. "I observed it," says he, "from sunrise till nine o'clock, again a little before ten, and lastly at noon, and from one to two o'clock; the rest of the day being devoted to higher duties, which might not be neglected for these pastimes."

3. The transit of 1769 was observed with intense interest by astronomers in both hemispheres. To secure the advantages of observations at different points, Capt. Cook was

---

117. Node months of Venus? Where line of nodes? Why June and December her node months? (Why only one month after those of Mercury?) 118. When last transit of Venus? Next? (How many have been observed? What said of Rittenhouse? Webster? When next four transits of Venus? When first transit noticed? What said of it? That of 1769 Cook—use of observations?)

sent to the Pacific in the bark "Endeavor," where he perished subsequently by the bands of savages at one of the Sandwich islands. Observations upon these transits furnish data for important astronomical calculations.

119. In consequence of the earth's annual revolution around the sun, he appears to travel *eastward*, through all the signs of the zodiac, every 365¼ days. It is this eastward motion of the sun that causes the stars to rise and set earlier and earlier every night.

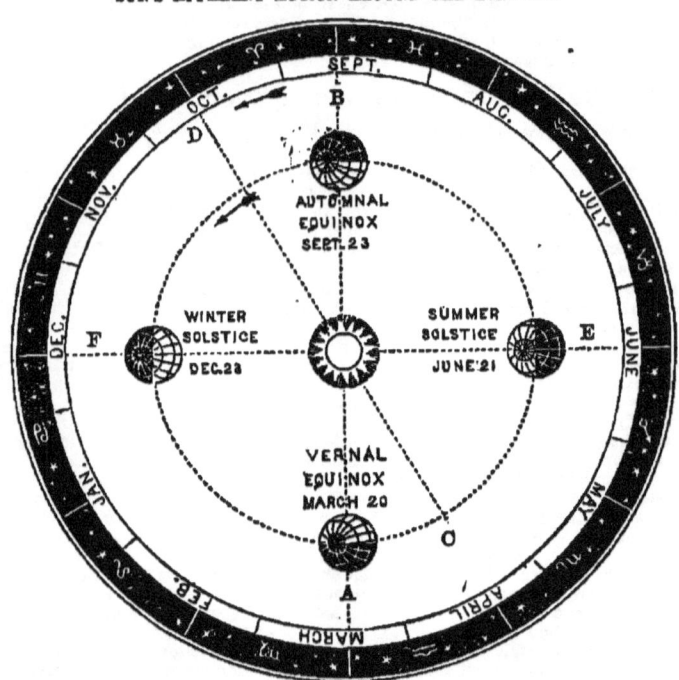

SUN'S APPARENT MOTION AROUND THE ECLIPTIC.

Let a person walk around a tree, for instance, at a short distance from it, and it will appear to sweep around the horizon in an opposite direction. So as the earth revolves annually about the sun, the sun appears to traverse the circle of the heavens in the opposite direction. Suppose the earth is at A on the 20th of March; the sun will appear to be at B in the opposite side of the ecliptic. As the earth moves on in her orbit from A to C, the sun will *appear* to move from B to D; and will seem thus to traverse the whole circle of the heavens every 365¼ days, or as often as the earth revolves around him. The *time* of the sun's apparent entrance into the different constellations, as he journeys eastward, is usually laid down in almanacs. Thus: "Sun enters ♈ (Aries) 20th of March, &c.;" at which time the *earth* would enter the sign ♒ (Aquarius), and the *sun* would seem to enter the opposite sign Aries.

119. What said of sun's apparent motion? Cause? Time of revolution? Effect upon the stars? (Illustration from tree? By diagram. What is meant by the sun's entering Aries? When? Where earth then /

CHAPTER II.

PRIMARY PLANETS CONTINUED.

120. BESIDES the revolution around the sun, the planets all revolve rapidly about their respective axes, as they perform their celestial journeys. This is called their *diurnal revolution*.

<small>The evidences of the earth's revolution have already been considered on pages 13 and 14. That most of the other planets revolve has been ascertained by carefully observing the motions of spots, as they seemed to pass periodically over their disks.</small>

121. The axis of the earth is inclined to the plane of the ecliptic 23° 28'. It is always *parallel to itself*—that is, it always inclines the *same way, and to the same amount*.

INCLINATION OF THE EARTH'S AXIS TO THE PLANE OF THE ECLIPTIC.

<small>1. The inclination of the earth's axis, and its parallelism to itself, are exhibited in the above cut, as also in the cuts, pages 50, 51, and 64, to which the student will do well to turn.
2. The author is aware that the poles of the earth have a slow motion around the pole of the ecliptic, requiring 25,000 years for a single revolution, but prefers to consider this point hereafter, in connection with the precession of the equinoxes.</small>

122. The axes of all the planets are inclined more or less to the planes of their respective orbits. This inclination, so far as known, is as follows:

Venus . . . 75°	Jupiter . . . 3° 04'
Mars . . . 28° 42'	Saturn . . . 26° 50'

<small>120. What revolution have the planets besides around the sun? What called? (What proof of the earth's revolution? Of the other planets?)
121. What said of the axis of the earth? Of the stability of its inclination? (Is there no variation?)
122. Are the axes of the other planets inclined? To what extent, respectively? (Substance of note 1? Illustrate by diagram. Note 2?)</small>

1. The student will bear in mind that the above inclination is not to the *ecliptic*, or plane of the earth's orbit, but to the plane of the *orbits* of the several planets respectively. Take the case of Venus, for instance:

The *orbit* of Venus departs from the *ecliptic* 8¼°, as stated at 108, while her *axis* is inclined to the plane of her orbit 75°, as shown in the above figures. This distinction should be kept definitely in view by the student.

2. The inclination of the axes of the several planets, each to the plane of its own orbit, is represented in the following cut:

INCLINATION OF THE AXES OF THE SEVERAL PLANETS TO THE PLANES OF THEIR ORBITS.

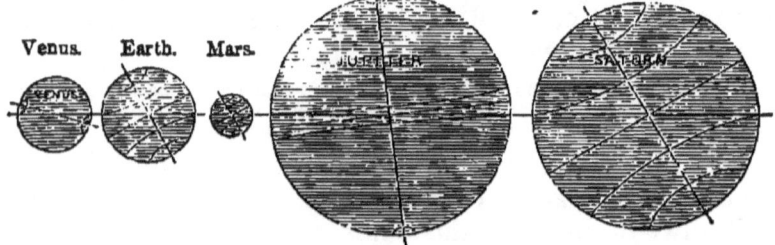

123. The inclination of the earth's axis to the plane of the ecliptic causes the equinoctial to depart 23° 28′ from the ecliptic. This angle made by the equinoctial and the ecliptic is called the *Obliquity of the Ecliptic.*

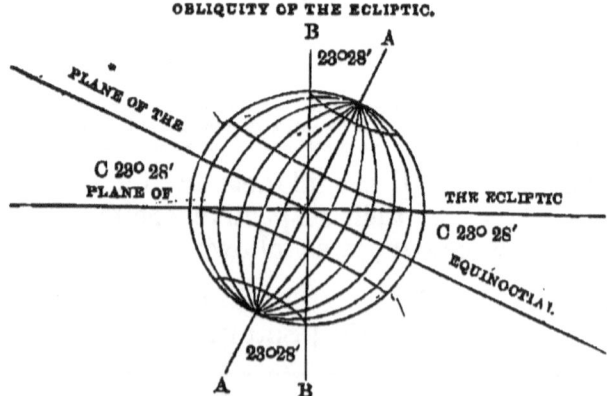

Let the line A A represent the axis of the earth, and B B the poles or axis of the ecliptic. Now if the line A A inclines toward the plane of the ecliptic, or, in other words, departs from the line B B to the amount of 23° 28′, it is obvious that the plane of the

123. What effect has the inclination of the earth's axis upon the equinoctial? What is the obliquity of the ecliptic? (Illustrate by diagram.)

equator, or equinoctial, will depart from the ecliptic to the same amount. This departure, shown by the angles C C, constitute the *obliquity of the ecliptic*.

124. The permanent inclination of the earth's axis, and her revolution around the sun, cause first one pole to be enlightened and then the other, thus producing the *seasons*. The same inclination and revolution cause the sun to appear to oscillate from north to south, crossing the equator twice every year. This is called the *sun's declination*. (See page 26.)

This subject of the seasons will be sufficiently understood by examining the cuts on pages 64 and 65.

125. The *equinoctial points* in the earth's orbit are two points in opposite sides of the ecliptic, at which the sun is exactly in the equinoctial; or, in other words, the plane of the equinoctial exactly cuts the sun's center. The first of these is passed on the 20th of March (the sun beginning then to decline northward), on account of which it is called the *vernal equinox;* and the other on the 23d of September, on account of which it is called the *autumnal equinox*. (See the earth at A and B, in the cut, page 64.)

If the sun is vertical at the equator, he will, of course, shine to both poles, as represented in the cut, and the days and nights will be equal all over the world. Hence the name *equinoctial*, from the Latin *æquus*, equal, and *nox*, night.

126. The *solstitial points* are those points in the earth's orbit where the sun ceases to decline from the equinoctial, and begins again to return toward it. They are respectively 90° from the equinoctial points.

The *Summer Solstice* is reached on the 21st of June, when the sun has the greatest *northern* declination, and it is summer in the northern hemisphere.

The *Winter Solstice* is reached on the 23d of December, when the sun has the greatest *southern* declination, and it is summer in the southern hemisphere, and winter in the northern. (See the earth at E F, cut, page 64.)

---

124. What other effects from the inclination of the earth's axis? Sun's declination?
125. What are the *equinoctial points?* How distinguished, and why? When passed? (Substance of note?)
126. The *solstitial* points? How far from the equinoctial points? How distinguished? When passed?

127. The amount of the sun's declination north and south of the equinoctial is 23° 28'; answering to the inclination of the earth's axis, by which it is caused, and marking the limits of the tropics upon the earth's surface.

1. On the 21st of June the sun reaches his greatest *northern* declination, or *Summer Solstice*, and is vertical on the Tropic of Cancer. From this time he approaches the equator of the heavens till the 20th of September, when he crosses it, and begins to decline *southward*. On the 23d of December he has reached his greatest *southern* declination, or *Winter Solstice*, and begins to return toward the equinoctial, which he passes on the 20th of March, and reaches his Summer Solstice again on the 21st of June. In this manner he continues to decline, first north and then south of the equator, from year to year. But it should not be forgotten that the sun does not really move, first north and then south, but that the apparent motion is caused simply by the inclination of the earth's axis and her revolution around the sun.

SHADOWS AT THE EQUATOR.

2. The sun's declination may be easily measured by the shadow of a suitable object upon the earth's surface. Suppose the flag-staff in the cut to stand perpendicularly, and exactly on the equator. On the 23d of December the shadow would be thrown *northward* to A, or 23° 28'—just as far as the sun has declined south. At 12 o'clock, on the 20th of March, and the 23d of September, there would be *no shadow;* and on the 21st of June, it would extend *southward* 23° 28' to C. Thus, at the equator, the shadow falls first north and then south of all perpendicular objects, for six months alternately.

MEASURING THE SUN'S DECLINATION IN NORTHERN LATITUDE.

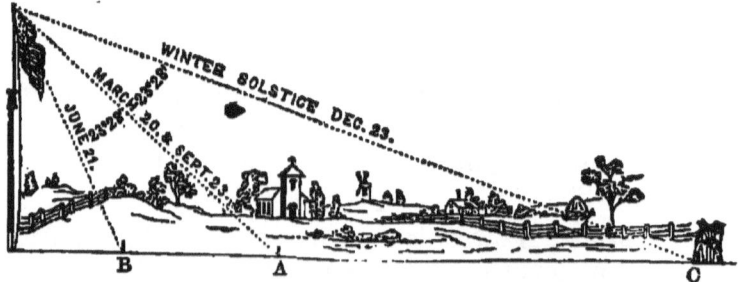

3. This cut shows how the student may measure the sun's declination wherever he may be located north of the equator. The shadows are such as are cast by objects during the year, about 45° north of the equator. On the 23d of December, when the sun has his greatest declination, the shadow of the flag-staff extends *north* at 12 o'clock to the point C, where two boys are seen, having just driven down a stake. From this time to June 21st the shadow gradually *shortens*, till on that day it reaches the point B, where another stake is driven. It then begins to elongate, and in six months is extended to C again. The point A is just half-way from B to C in *angular* measurement, though the distances on the plain in the picture are very different. When the sun is on the equator, March 21st and September 23d, the shadow will reach only to A; and the angle A B and the top of the staff shows the *northern*, and A C and the top of the staff the *southern* declination. It will be found to be 23° 28' each way, as marked in the figure.

---

127. To what extent does the sun decline from the equinoctial north and south? Why not more? (Substance of note 1? Note 2, and explain by diagram. Note 3, and diagram. What is a *gnomon?*)

4. The angle formed by the top and bottom of the pole and the point A will exactly correspond with the latitude of the place where the experiment is made.

5. Let the students try this matter for themselves. Select a level spot, and put up a stake, say ten feet high. Get an exact "noon mark," or north and south line, where the stake is driven, and at 12 o'clock, every fair day, put down a small stake at the end of the shadow. In this manner you will soon be able to measure the sun's declination for yourselves, to determine the latitude of the place where you live, and to understand how mariners at sea ascertain their latitude by the declination of the sun.

6. The ancients had pillars erected for the purpose of making observations upon their shadows. Such a pillar is called a *gnomon*.

ROTATION OF THE PLANETS UPON THEIR AXES.

128. The time, so far as known, of the revolution of the planets upon their respective axes, or, in other words, the length of their natural days, is as follows:

	h.	m.		h.	m.
Mercury	24	5	Juno	27	0
Venus	23	21	Jupiter	9	56
Earth	24	00	Saturn	10	29
Mars	24	37	Uranus	9	30

These statistics are given upon the authority of Sir John F. W. Herschel, though he marks Juno and Uranus as doubtful.

129. The revolution of the earth upon its axis is the cause of the agreeable vicissitudes of day and night.

PHILOSOPHY OF DAY AND NIGHT.

 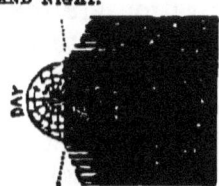

How wisely adapted to the happiness of His creatures are all the works of God! The night prepares us for the day, and the day in turn prepares us to welcome the night; and in both instances the change ministers to the happiness of man and beast. And but for being carried around into the darkness of the earth's shadow, we should never have admired the dazzling firmament, as it declared the glory of God, and showed forth his handiwork. How beautiful the poetic allusion to the revealing power of night!

> Mysterious Night! when our first parent knew
> Thee, from report divine, and heard thy name,
> Did he not tremble for this lovely frame,
> This glorious canopy of light and blue?
> Yet, 'neath a curtain of translucent dew,
> Bathed in the rays of the great setting flame,
> Hesperus with the host of heaven came;
> And lo! creation widen'd in men's view.

---

128. In what time do the other planets rotate on their respective axes? (Note?)

129. Cause of day and night? (Substance of note? Poetic quotation?)

> Who could have thought such darkness lay conceal'd
> Within thy beams, O Sun! or who could find,
> Whilst fly, and leaf, and insect stood revealed,
> That to such countless orbs thou mad'st us blind?
> Why do we, then, shun death with anxious strife?
> If light can thus deceive us, may not life?

130. The earth and all the other planets revolve *eastward* upon their axes, or in the same direction that they revolve in their orbits. This also is determined (with the exception of the earth) by observing the motion of *spots* upon their surfaces, by the aid of telescopes.

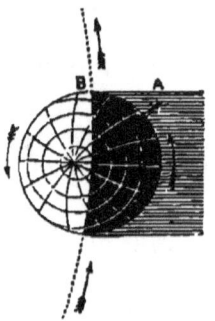

DIURNAL REVOLUTION OF THE PLANETS.

1. In the cut we have an arc of the earth's orbit, and the earth revolving on her axis as she revolves around the sun. The *arrows* show the direction in both cases.
2. By holding the book up south of him, and looking attentively at the cut, the student will understand why the sun "rises" or first appears in the *east*. It is because the earth revolves eastward. Thus the observer at A is carried round into the light, and sees the sun rise when he reaches B.

TIME.

131. *Time* is duration measured either by natural or artificial means. The principal *natural* indicators of the lapse of duration are the revolution of the earth upon its axis, marking a natural day; the change of the moon, denoting a lunar month; and the cycle of the seasons, denoting a year. Time is measured *artificially* by clocks, watches, chronometers, dials, &c.; the standard being the solar day still, which is divided artificially into 24 parts, called *hours*, and these again into *minutes* and *seconds*.

The aboriginal tribes of this country all reckoned time by "moons," or months, as denoted by the moon's changes.

132. The motion of the earth upon its axis is the most regular of which we have any knowledge. It does not vary one second in a thousand years.

To this stability of the earth's motion upon her axis the prophet refers when he says: Thus saith the Lord, If ye can break my covenant of the day, and my covenant of the

---

130. In what direction do the planets rotate on their axes? How ascertained? (Explain why the sun appears to *rise* in the east.)
131. What is *time?* What natural standards? Artificial? (How measured by aborigines?)
132. What said of earth's motion on axis? (What reference to in Scriptures?)

night, and that there should not be day and night in their seasons, then may also my covenant be broken with David," &c.—Jeremiah xxxiii. 20.

133. Time is of two kinds—*Solar* and *Sidereal*. A *solar* day is the time elapsing from the sun's crossing the meridian of any place, to his coming to the same meridian again. A *sidereal* day is the time intervening between the transit of a *star* across the meridian, to its coming to the same meridian again.

134. A solar day consists of 24 hours, at a mean rate, but a sidereal day is accomplished in 23 hours, 56 minutes, and 4 seconds; the solar day being near 4 minutes the longest. This slight difference of about 4 minutes daily, between solar and sidereal time, amounts to *one whole day* in every $365\frac{1}{4}$. Owing to the revolution of the earth around the sun, and his apparent annual revolution eastward among the stars, it requires 366 revolutions of the earth, as measured by the fixed stars, to make $365\frac{1}{4}$ days, as measured by the sun.

135. The *cause* of this difference in the apparent revolutions of the sun and stars, and consequent difference in the length of a natural day, as measured by the passage of a *star* or of the *sun* across the meridian, is this: The earth is constantly advancing in her orbit while she revolves on her axis, causing the sun to *appear* to move slowly eastward among the stars; or, what is the same thing, the stars to appear to rise earlier and earlier every night, and one after another to overtake and pass by the sun. (See Article 119.) When, therefore, the meridian is brought around to that point in the heavens where the sun was near 24 hours before, he is not there, but has moved a little *eastward*. But a star that, 24 hours before, was exactly behind the center of the sun in the distant heavens, will be found *west* of the sun, and will consequently cross the meridian before the sun does. The time required for the meridian to revolve from the star to the sun constitutes the 3 minutes 56 seconds difference between solar and sidereal time.

133. Kinds of time? Define each.
134. Length of solar day? Sidereal? Difference? Amount in year?
135. State the *cause* of the difference in the time of the apparent revolution of the sun and stars. Illustrate by diagram.

ASTRONOMY.

SOLAR AND SIDEREAL TIME.

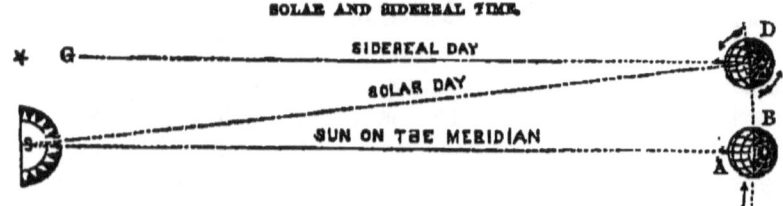

1. To the man at A the sun (S) is exactly on the meridian, or it is twelve o'clock, noon. The earth passes on from B to D, and at the same time revolves on her axis. When she reaches D, the man who has stood on the same meridian has made a complete revolution, as determined by the star G (which was also on his meridian at twelve o'clock the day before); but the sun is now *east* of the meridian, and he must wait *four minutes* for the earth to roll a little further eastward, and bring the sun again over his north and south line. If the earth was not revolving around the sun, her solar and sidereal days would be the same; but as it is, she has to perform a little more than one complete revolution each solar day, to bring the sun on the meridian.

EQUATION OF TIME.

136. As the distant stars have no motion, real or apparent, around the ecliptic, and the earth's motion upon it is uniform, it results that sidereal time is *always exactly the same*.

A clock that keeps sidereal time is called a *sidereal clock*. One of these instruments is almost indispensable in the observatory of the astronomer.

137. Solar time is constantly varying. No two successive solar days are exactly of a length. The 24 hours given as the length of a solar day (134) is the *average* of all the solar days throughout the year. Hence it is called *mean solar time*. The time, as indicated by the transit of the sun across the meridian, from day to day, is called *apparent time*.

138. A well-regulated clock will keep *mean solar time*, and will vary from the *apparent* time (as indicated by a noon mark, or dial) to the amount of $16\frac{1}{4}$ minutes one way, and $14\frac{1}{2}$ the other. The sun will at one time cross the meridian $16\frac{1}{4}$ minutes before it is noon by the clock—the apparent time being $16\frac{1}{4}$ minutes faster than mean or clock time; while at another time it will be noon by the clock $14\frac{1}{2}$ minutes before it is noon by the sun.

---

136. Is sidereal time always the same? Why must it be? (What is a sidereal clock?)
137. What said of the variations of solar time? What is *mean solar time*? Apparent?
138. What time do common clocks keep? How much variation from sun? How?

EQUATION OF TIME.

139. The difference between apparent and mean solar time is called the *Equation of Time*. It is greatest about the 3d of November, when the clock is 16 minutes and 17 seconds behind the sun. Four times a year—viz., April 15th, June 15th, September 1st, and December 23d—the clock and sun will agree; or, in other words, mean and apparent time will be alike.

140. The inequality of the solar days depends upon two causes—the unequal velocity of the earth in her orbit (77, 78), and the inclination of her axis to the plane of her orbit (123).

141. If the earth's orbit were an exact circle, she would move with the same velocity in all parts of it; and if she revolved with regularity upon her axis, her solar days would be exactly of a length.

EQUAL SOLAR DAYS.

Let the circle in the adjoining cut represent the earth's orbit, and the projections from the earth toward the sun a *pillar* or *gnomon* standing upon a given meridian. The cut will then show that with a circular orbit, and uniform motion in it, and a regular rotation upon her axis, the earth would bring the gnomon around toward the sun at regular intervals, both of distance in her orbit, and of time. In that case, all apparent solar days would be equal.

142. As the orbit of the earth is elliptical, it requires more time for the earth to pass from the vernal equinox, through the aphelion, to the autumnal equinox, than it does from the autumnal equinox, through the perihelion, to the vernal equinox. The difference is about eight days—the sun being north of the equinoctial about eight days longer than he is south of it. Hence the summers of the northern hemisphere are longer than the winters.

143. As the earth's orbit is an ellipse, and the earth

---

139. What this difference called? When greatest? When no difference?
140. What causes the inequality in the length of the solar days?
141. What necessary in order that they may be equal? (Illustrate by diagram and explanations.)
142. What effect has the ellipticity of the earth's orbit upon the length of the seasons, north and south of the equator?

moves faster in some parts of it than in others, while its rotary motion is uniform, it follows that its orbital velocity in longitude must sometimes be *faster*, and at others *slower* than its orbital motion, thus causing an inequality in the length of the solar days.

UNEQUAL SOLAR DAYS.

From A to B in the adjoining cut, the orbital motion is slower than its mean rate, and the rotary motion gains upon it. Hence the gnomon is shown revolving too fast, and as pointing east of the sun, when the earth has performed her journey for a mean solar day. From B to A, the earth's motion in her orbit gains upon her rotary motion, and the gnomon is behind, or *west* of the sun. At A and B the clock and sun would agree. From A to D the sun gains on the clock, till it gets 14¼ minutes ahead. From D to B this difference is diminished, till at B the sun and clock agree. From B to C the clock gains on the sun, till the difference is 16¼ minutes; and from C to A this difference diminishes, till at A mean and apparent time agree again.

144. The earth's perihelion is in ♊, and her aphelion in ♐; the first of which she passes on the first of January, and the latter on the 3d of July. We are consequently about three millions of miles nearer the sun Jan. 1, than July 3d.

The natural effect of this variation would be, so far as it had any influence, to modify the cold and heat in the Northern Hemisphere, and to augment both in the Southern. For instance, our nearness to the sun in January would slightly soften our winter, while, at the same time, it slightly increased the heat of the summer south of the equator. So, also, our increased distance in July would diminish the heat of our summer, and at the same time enhance the cold of the corresponding winter in the Southern Hemisphere. But the variation of 3,000,000 miles is so slight, when compared with the whole distance of the sun, that the change of temperature produced thereby is imperceptible.

THE CALENDAR, LEAP YEAR, OLD AND NEW STYLE, ETC.

145. The Julian calendar divided the year into 12 months, containing in all 365 days. But a full astronomical year, or the time requisite for the earth to revolve from one equinox around to the same equinox again, consists of 365d. 5h. 48m. 51s. Hence the Julian

---

143. Explain the cause of this inequality? (Illustrate by diagram.)
144. Where are the perihelion and aphelion points, and when passed? When nearest, and how much? (What effect?)
145. Describe the Julian calendar? An astronomical year? What difference? What effect? How corrected?

year was near 6 hours, or one day in every four years, too short; which, if left uncorrected, would in time completely reverse the seasons, giving harvests in January, and snow in July. To prevent this constant falling behind, a correction was applied, by adding one day to February every fourth year. Hence it is called *Bissextile* or *Leap Year*.

146. But one whole day added for every four years was 44m. 36s. *too much*. From A. D. 325 to 1582 this excess amounted to about 10 days; so that the civil year was thus much ahead of the astronomical. In 1582, Pope Gregory XIII. applied a further correction, or *reformed* the Julian calendar. To make the civil and astronomical years agree, so that the vernal equinox would happen on the 21st of March, as it did 1257 years before, Gregory resolved to strike out of the civil year the 10 days it had gained, and ordered that the 5th of October should be called the 15th. This reformed or corrected calendar is called the *Gregorian* calendar.

147. To prevent the civil year from running ahead of the astronomical again, in the lapse of centuries, by the 11m. 12s. which it exceeded the astronomical, it was prescribed that at certain convenient periods the intercalary day of the Julian period should be omitted. Thus the centennial years 1700, 1800, 1900, are, according to the Julian calendar, bissextiles; but on these it was ordered that the intercalary day should not be inserted, inserted again in 2000, but not inserted in 2100, 2200, 2300; and so on for succeeding centuries.

148. The Gregorian or reformed calendar was adopted as soon as promulgated, in all Catholic countries; but in England, the "change of style," as it was called, did not take place till September, 1752. Eleven nominal days were then struck out, and the 3d of September was called the 14th. At the same time, the time of the beginning

---

146. Was the calendar then correct? Why not? What result? Who corrected? When? How? What this reformed calendar called?
147. What further correction necessary? How effected?
148. Was the Gregorian calendar at once adopted? When in England? How then adopted? What other change at the same time? What effect in

of the civil year was changed from the 25th of March to January 1st, as it now stands. The year 1752, which was to have begun on the 25th of March, was made to begin on the 1st of January preceding; so that for dates falling between the 1st of January and the 25th of March, the number of the year is one greater by the *New* than by the *Old Style*. And as the intercalary day was omitted in 1800, there is now, for all dates, 12 days difference between the old and new styles. Russia is now the only Christian country in which the Gregorian calendar is not used.

TIME, AS AFFECTED BY LONGITUDE.

149. As the sun's crossing the meridian of any place determines it to be 12 o'clock, apparent solar time, at that place, it is evident that 12 o'clock comes *sooner* to places east on the earth's surface, and *later* to places west.

1. Let the adjoining cut represent the earth, the arrows indicating the direction of her revolution, and the sun being on the meridian at XII. at the top. It will then be day over all the light portion of the globe, and night over all the shaded portion. On the meridian exactly under the sun it is just XII. o'clock noon; while at the meridian on the opposite side of the earth it is just 12 o'clock at night, or midnight. When the light and shade meet on the right, it is VI. o'clock morning; and directly opposite on the left, is VI. o'clock evening.

2. Observe that when it is XII. at A, it is I. o'clock at B, II. o'clock at C, &c., while it is only XI. o'clock at D, X. o'clock at E, IX. o'clock at F, &c.; thus showing how it is that time is earlier east, and later west of any given meridian.

150. Every 15° of longitude upon the earth's surface makes an hour's difference in the time. If *east* of the given meridian, it will be an hour *earlier;* if *west*, an hour *later*.

---

reckoning years of time? What the difference now between *Old* and *New* style, and why? What calendar used in Russia?

149. What effect has the longitude of a place upon its time? (Diagram, and explain?)

150. What difference of longitude is required to make an hour's difference in time? When *earlier?* When *later?* (How demonstrated? When 6

1. If the sun passes through 360° every 24 hours, he must pass over 15° each hour, as 360° ÷ 24 = 15°. Hence every 15° must make an hour's difference in the time; and when it is sunrise, or 6 o'clock, solar time, in New York city, it will be noon, or 12 o'clock, 90° east of New York, and midnight 90° west of it.

2. Taking the circumference of the earth at 25,000 miles, the sun passes over $1041\tfrac{2}{3}$ miles every hour at the equator; for 25,000 miles ÷ 24 equals $1041\tfrac{2}{3}$ miles. And if $1041\tfrac{2}{3}$ miles be divided by 60, the number of minutes in an hour, it gives about $17\tfrac{1}{4}$ miles as the space over which the sun travels at the equator every minute. Every $17\tfrac{1}{4}$ miles, therefore, east or west, will make one minute's difference in the time. As we recede from the equator north or south, the meridians approach each other, and a degree of longitude becomes less and less to the poles.

3. A person leaving Boston with the exact time will find, on reaching Albany, about 3° west of Boston, that his watch is some 12 minutes ahead of the Albany time; and on reaching Buffalo, about 5° further west, that it is some 32 minutes ahead of the true time at Buffalo. So in traveling from Buffalo to Boston, the Albany and Boston time will be found to be the same extent *ahead* of the Buffalo time. Hence conductors on railroads, running their trains by time, set their watches from Albany to Buffalo by some standard agreed upon—as, for instance, Syracuse time—and reject all other local time, be it faster or slower.

151. As every 15° upon the earth's surface makes an hour's difference in the time, it is easy to convert degrees into time, or time into degrees. By this means, a mariner having the time at the place whence he sailed, and the time where he is, from observing when the sun crosses the meridian, can ascertain, from the *difference* between his standard and local time, his distance east or west of the port whence he sailed, or, in other words, his longitude.

1. Time is converted into degrees by multiplying the hours by 15 for the degrees, and adding one-fourth of the minutes to the product; for every minute of time makes $\tfrac{1}{4}$°, and every second of time $\tfrac{1}{4}$' in longitude.

2. On the other hand, degrees of longitude are converted into time by *dividing* them by 15 for the hours, and multiplying the remainder, if any, by 4 for the minutes, &c.

152. The rotation of the planets upon their respective axes has caused them to swell out at their equators, and contract at their poles—thus assuming the form of oblate spheroids (page 18).

1. When fluids are left free to yield to the influence of attraction, as mutually existing between their particles, they invariably assume a *spherical form*. Hence water, in falling from the clouds, takes the form of spherical drops; and melted lead, thrown from the top of a shot-tower, takes a spherical form, and cooling in the air on its passage down, remains perfect little globes, called *shot*.

2. A *solid* sphere would never become oblate by revolution. It might burst, from its powerful centrifugal tendency, as grindstones sometimes do in manufactories of cutlery; but it must be *fluid*, or at least soft and yielding, in order to become oblate by revolution.

o'clock in New York, what time 90° east?—90° west? How many miles does the sun pass over in an hour at the equator? Per minute? How determined? How north and south of equator? At 45th degree? What difference from Boston to Albany and Buffalo? From Buffalo to Boston? Hence what practice?)

151. Can time be converted into degrees, and degrees into time? How useful in navigation? (How convert time into degrees? Degrees into time?)

152. Effect of rotation upon figure of planets? (Note 1? 2. Solids? 3. What does oblateness indicate? 4. Proof from Scriptures? Remark?)

3. The oblateness of the planets, then, seems to indicate two things: First, that they were all once in a fluid or plastic state; and, secondly, that they began to revolve while in that state, or before any part of them had become solid, like our continents and islands.

4. So far as the earth is concerned, we are taught in the Holy Scriptures—the best and most accurate of all books—that the earth and water of our globe were once so mixed, that the whole appeared as a "void" of "waters;" and that they were afterward separated into "earth" and "seas" by the Almighty Creator. (See Genesis i., 2, 9, 10.) Thus we see that true science and the Bible are always in harmony with each other.

153. The difference between the polar and equatorial diameters of the planets, so far as known, is as follows: the Earth, 26 miles; Mars, 25; Jupiter, 6,000; and Saturn, 7,500.

The oblateness of Jupiter and Saturn is as plainly visible through a telescope, as the difference in the following figures is to the eye of the student.

ORIGINAL FORM.   CHANGE.   PRESENT APPEARANCE.

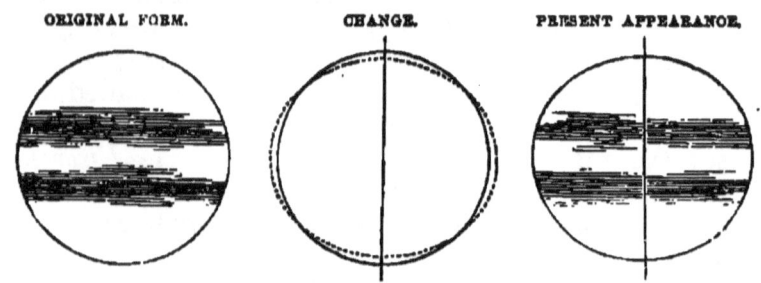

The plain line in the middle figure shows the original form, and the dotted line its present form. The difference is the change produced by its rotation. When measured by the proper instruments, it is found, in the case of Jupiter, to amount to about $\frac{1}{15}$ of his average diameter; and that being 89,000 miles, $\frac{1}{15}$ is but little less than 6,000.

154. As Mercury and Venus rotate in about the same time of our globe, and their sidereal years are only 88 and 225 days respectively (72), it follows that Mercury has but 88 natural days to his year, and Venus only about 225 to hers. But the natural day of Jupiter being only 10 hours long, and his year equal to about 12 of ours (11 years 317 days), he must have 10,397 natural days in one of his years. So Saturn's year, consisting of 29 years 175 days of our time, will allow him to rotate on his axis about 25,000 times; or, in other words, will allow of 25,000 natural days in each of his years. The year of Uranus being equal to 84 years and 27 days of

---

153. State the difference of equatorial and polar diameters of planets? (Remark respecting Jupiter and Saturn?)

154. How many natural days has Mercury in his year? Venus? Jupiter? How so many? Saturn? Uranus? (Demonstrate.)

our time (71), and his diurnal revolution 9½ hours (128), it follows that he has 92,683 natural days in his year.

<small>29 years 175 days = 10,760 days of our time; ×24 = 258,240 hours ÷ 10¼ hours, the time of Saturn's revolution, = 24,594 $\frac{3}{10}$, the number of days in his year. So 84 years, 27 days, the periodic time of Uranus = 36,657 days, or 880,488 hours; which ÷ 9¼ hours, the time of the planet's diurnal revolution = 92,683, the number of natural days in his year.</small>

155. As going *from* the earth's center is to *ascend* (page 27), and the equator of an oblate spheroid is further from the center than the poles, it follows, that the earth being an oblate spheroid, we must ascend somewhat in going from either pole to the equator. A river, therefore, running for a great distance toward the equator, would actually *ascend;* or, in other words, run up hill —the centrifugal force generated by the earth's motion driving the water on toward the equator.

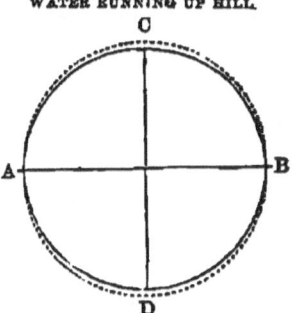

WATER RUNNING UP HILL.

<small>The Mississippi is said to be higher at its mouth than it is some thousands of miles north of it. If its bed conforms at all to the general figure of the earth, this must certainly be the case, as may be demonstrated by the aid of the annexed diagram. Let A B represent the polar, and C D the equatorial diameters. The entire difference between them is 26 miles, or 13 miles on each side. The two circles represent this difference. Now as the earth's circumference is 25,000 miles, the distance from the poles to the equator (being one-fourth of that distance) must be 6,250 miles; and in that 6,250 miles the ascent is 13 miles, or over two miles to every 1,000 toward the equator. The Mississippi runs from the 50th to the 30th degree of north latitude inclusive, or 21 degrees; which, at 69¼ miles to a degree, would amount to about 1,500 miles. If, then, it runs a distance equivalent to 1,500 miles directly south (in a winding course of about 8,000), theory requires that it should be about three miles higher at its mouth than it is 1,500 miles directly north. There is some philosophy, therefore, in saying that if a river runs for a great distance from either pole toward the equator, it must run up hill.</small>

156. Should the earth cease to rotate upon its axis, the waters about the equator would at once rush toward the poles, flooding them to the depth of 6½ miles, and receding from the equator to the same amount. So far as the solid portions of the earth would permit, it would at once become a perfect sphere. (See page 17, and also Art. 153 and note.)

157. It has already been stated (77), that the orbits of all the planets were ellipses; but they are not all alike eccentric. The orbit of Mercury is quite elliptical, while

<small>155. What curious fact follows from the earth's oblateness? (What instance given? Illustrate by diagram.)
156. What would be the effect should the earth cease to rotate?</small>

that of Venus is nearly a circle. The student should observe that the eccentricity is not the deviation from a circle, but the distance from the center of an ellipse to either foci (see page 23 and cuts).

The eccentricity of the orbits of the principal planets is as follows:

	Miles.		Miles.
Mercury	7,000,000	Ceres	21,000,000
Venus	492,000	Pallas	64,250,000
Earth	1,618,000	Jupiter	24,000,000
Mars	13,500,000	Saturn	49,000,000
Vesta	21,000,000	Uranus	85,000,000
Astræa	—	Neptune	—
Juno	64,000,000		

PRECESSION OF THE EQUINOXES.

158. The equinoctial points have already been defined (125) as two points in the earth's orbit where the equinoctial or celestial equator (20) cuts the sun's center. They are in opposite sides of the ecliptic, or 180° apart (see 119 and cut). The vernal equinox is the point from which both celestial longitude and right ascension are reckoned (20 and 91); but not being marked by any fixed object in the heavens, it is reached just when the sun comes to be exactly over the earth's equator, or in the equinoctial

159. But it is found by long and careful observation that the earth reaches the equinoctial point about 22 minutes and 23 seconds earlier every year than on the year preceding. This is equal to $50\frac{1}{4}''$ of arc in the ecliptic. In this manner the equinoctial points are slowly receding westward,

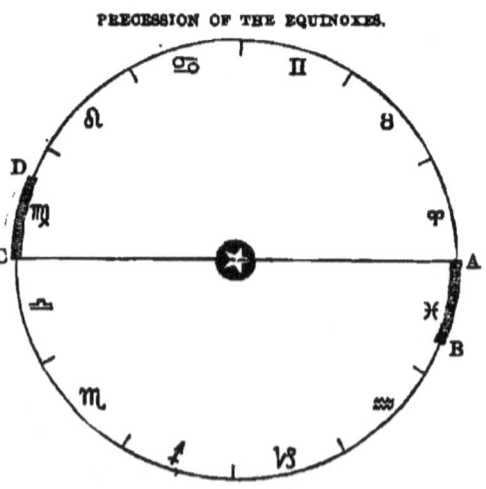

PRECESSION OF THE EQUINOXES.

---

157. What said of the orbits of Mercury and Venus? Of eccentricity?
158. Are the equinoctial points marked by any fixed object in the heavens? How know when reached?
159. Are they stationary or not? Reached how much earlier annually?

or falling back upon the ecliptic, at the rate of 50¼″ a year, or 1° every 71⅔ years. This would amount to 30°, or one whole sign in 2,140 years, and to the entire circle of the ecliptic in 25,868 years.

This very interesting phenomenon may be explained by the preceding diagram. Let the point A represent the vernal equinox, reached, for instance, at 12 o'clock on the 20th of March. The next year the sun will be in the equinoctial 22 minutes 28 seconds earlier, at which time the earth will be 50¼″ on the ecliptic, back of the point where the sun was in the equinoctial the year before. The next year the same will occur again; and thus the equinoctial point will recede westward little by little, as shown by the small lines from A to B, and from C to D. It is in reference to the stars going forward, or seeming to *precede* the equinoxes, that the phenomenon was called the Precession of the Equinoxes. But in reference to the motion of the equinoxes themselves, it is rather a *recession*.

160. The *cause* of this wonderful motion was unknown, until Newton proved that it was a necessary consequence of the rotation of the earth, combined with its elliptical figure, and the unequal attraction of the sun and moon on its polar and equatorial regions. There being more matter about the earth's equator than at the poles, the former is more strongly attracted than the latter, which causes a slight gyratory or wabbling motion of the poles of the earth around those of the ecliptic, like the pin of a top about its center of motion, when it spins a little obliquely to the base.

161. One marked effect of this recession of the equinoxes is an increase of longitude in all the heavenly bodies. As the vernal equinox is the zero or starting point, if that recedes westward, it increases the distance between it and all bodies east of it; or, in other words, increases their longitude to the amount of its recession. Hence catalogues of stars, and maps, showing their longitude, need to be corrected at least every 50 years, otherwise their longitude, as laid down, will be too little to indicate their true position. Allowing the world to have stood at this date (1853) 5,857 years, the equinoxes have receded already through about 75° of longitude. At the same time the *constellations* have gone forward

---

How much in angular measurement? Revolving which way? At what rate? How long for 1°? For 30°? For the whole circle of the ecliptic? (Illustrate by diagram.)
160. *Cause* of recession? Who discovered?
161. Effect of recession upon longitude? Explain how effected. Signs and constellations?

eastward, and left the *signs* which bear their names. Hence the *sign* Aries actually covers the *constellation* Pisces.

162. Another effect of the recession of the equinoxes is, that it gives to the pole of the earth a corresponding revolution around the pole of the ecliptic in 25,868 years.

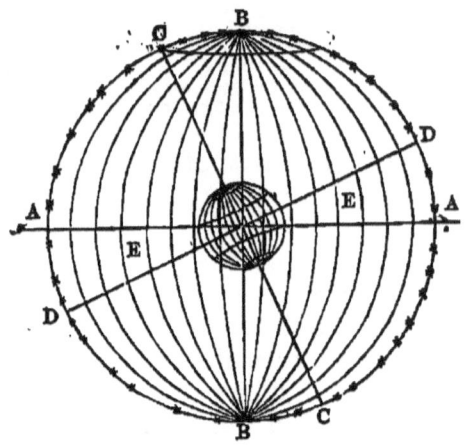

Let the line A A in the figure represent the plane of the ecliptic; B B, the poles of the ecliptic; C C, the poles of the earth; and D D, the equinoctial. E E is the obliquity of the ecliptic. The star C at the top represents the pole star, and the curve line passing to the right from it may represent the circular orbit of the north pole of the heavens around the north pole of the ecliptic.

163. This gyratory motion of the north pole of the heavens, while it keeps at the distance of 23° 28′ from the pole of the ecliptic, will cause it to change its place in the heavens to the amount of 46° 56′ in 12,934 years; thus alternately approaching toward and receding from the stars, at every revolution of the equinoxes around the ecliptic. Thus the place of the pole is in constant but very slow motion around the pole of the ecliptic.

164. The *Nutation* of the earth's axis is another small and slow gyratory motion, by which, if subsisting alone, the pole would describe among the stars, in the period of about 19 years, a minute ellipse, having its longer axis equal to 18″, and its shorter about 14″; the longer axis pointing toward the pole of the ecliptic. It is on account of these varied motions shifting the point from which longitude and right ascension are reckoned, and also the pole of the heavens, that it becomes necessary, in de-

162. What other effect of recession? (Illustrate by diagram.)
163. What effect upon the apparent distance of the stars from the north pole of the heavens?
164. What is *Nutation?* What meant by *epoch*, and why necessary to state?

scribing the place of a star or planet, by any of these standards, to state the *epoch* or time, and also whether it be *mean* right ascension—*i. e.*, right ascension after having been corrected for the recession of the equinox, the zero point.

165. The *Colures* are two great circles crossing at the *poles of the ecliptic* at right angles. One passes through the equinoxes, and is thence called the *Equinoctial Colure;* the other passes through the solstices, and is called the *Solstitial Colure*. They are to the heavens what four meridians, each 90° apart, would be to the earth.

CHAPTER III.

TELESCOPIC VIEWS OF THE PLANETS.

166. By the aid of telescopes, we discover myriads of objects in the heavens that are entirely invisible to the naked eye; while objects naturally visible are immensely magnified, and seem to be brought much nearer the observer.

This impression of nearness is an intellectual conclusion drawn from the fact of the increased distinctness of the object; as we judge of the distance of objects, in a great measure, by their dimness or distinctness.

MERCURY.

167. Under favorable circumstances, Mercury is visible to the naked eye, but yet is seldom seen, owing to his nearness to the sun. During a few days in March and April, and August and September, he may be seen for several minutes in the morning or evening twilight, when

---

165. What are the *colures?* Describe.
166. Effect of the telescope upon vision? Upon distant objects? (Why appear nearer?)
167. Can Mercury be seen by the naked eye? Is he often seen? Why not? When may he be seen? How appear?

his greatest elongations (99) happen in those months. He appears like a star of the third magnitude, with a pale rosy light. See 104 and note.

168. Through a telescope, Mercury exhibits different *phases* in different parts of his orbit, similar to those presented by the moon in her revolution around the earth. The German astronomer, Schröeter, discovered numerous mountains upon the surface of Mercury, one of which he estimated to be nearly 11 miles in hight. By observing these at different times, he determined the diurnal revolution of the planet to be 24h. 5m. 28s. But these observations have not been confirmed by any other astronomer. The apparent angular diameter of Mercury varies from 5" to 12", according to his position with respect to the earth (56 and 80). So far as is known, he is not attended by any satellite.

VENUS.

169. When favorably situated, Venus is one of the most conspicuous members of the planetary system, and is a most brilliant object even to the naked eye. Her color is of a silvery white, and, when at a distance from the sun, either east or west, she is exceedingly bright and beautiful. When nearest the earth, her apparent diameter is 61", which is greater than that of any other planet, owing to her being so much nearer than Jupiter or Saturn.

Under a telescope, Venus exhibits all the phases of the moon, as she revolves around the sun. The *cause* of this phenomenon is, that we see more of her enlightened side at one time than at another; and the same is true of Mercury.

1. The telescopic appearance of Venus, at different points in her orbit, is represented in the following figure. At E and W she has her greatest eastern and western elonga-

---

168. How appear through telescope? What said of Schröeter? What conclusion from observing the spots? Confirmed by others, or not? Angular diameter of Mercury? Why vary? Has he a satellite?
169. What said of Venus? Her apparent diameter? Why greater than that of Jupiter? How appear through telescope? *Cause* of her phases? (Describe phases when *east* of the sun—west. What prediction before the discovery of the telescope?)

tion, and is stationary; while her positions opposite the words "direct" and "retrograde" represent her at her conjunctions. The spots on the face of the sun represent Venus projected upon his disk, in a transit, the arrow indicating her direction.

TELESCOPIC PHASES OF VENUS.

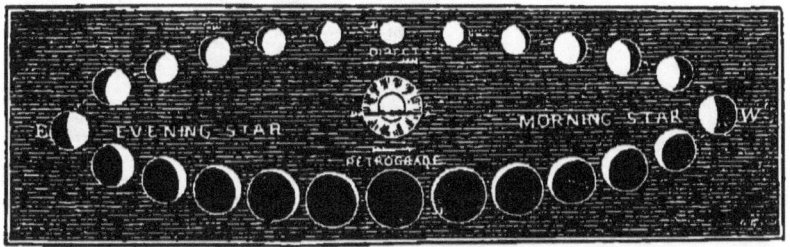

2. Before the discovery of the telescope, it was asserted that if the Copernican theory were true, Mercury and Venus would exhibit different phases at different times; and as these phases could not be seen, it was evident that the theory was false. But no sooner had Galileo directed his small telescopes to these objects, than he found them exhibiting the very appearances required by the Copernican theory, its opponents themselves being judges.

170. Besides the phases above mentioned, a close inspection of Venus will reveal a variety of *spots* upon her surface. These are supposed to be the natural divisions of her surface, as continents, islands, &c. Schröeter measured several *mountains* upon this planet, one of which he estimated at over twenty miles in hight. There is evidence of the existence of an *atmosphere* about this planet, extending to the distance of about three miles.

SPOTS SEEN UPON THE SURFACE OF VENUS.

171. Were a person situated upon one of the exterior planets, at a distance from our globe, it would exhibit phases like Mercury and Venus, in its annual revolution; and the continents, islands, and seas would appear only as *spots* upon her surface, assuming various forms, according to the position from which they were viewed.

---

170. What else seen upon Venus? What supposed to be? Schröeter's measurements? Has Venus an *atmosphere?*
171. How would our globe appear if viewed from a distance?

ASTRONOMY.

DISTANT TELESCOPIC VIEWS OF THE EARTH.

Above we have four different views of our own globe. No. 1 is a view of the Northern Hemisphere; No. 2, of the Southern; No. 3, of the Eastern Continent; No. 4, of the Western. A common terrestrial globe will present a different aspect from every new position from which it is viewed; as the earth must in her appearance to the inhabitants of other worlds.

MARS.

172. Mars usually appears like a star of the second magnitude, of a reddish hue. When in opposition, or nearest to the earth, he appears quite brilliant, as we see his disk fully illuminated. His apparent diameter is then about 18''; whereas, when on the opposite side of the ecliptic, or in conjunction with the sun (80), it is only 4''. He exhibits slight phases, and his surface seems to be variegated with hill and vale, like the other planetary bodies. "Upon this planet," says Dr. Herschel, "we discern, with perfect distinctness, the outlines of what may be continents and seas." When it is winter at his north pole, that part of the planet is *white*, as if covered with ice and snow; but as summer returns to his northern hemisphere, the brightness about his north pole disappears.

173. The general *ruddy color* of Mars is supposed by Sir John Herschel to indicate "an ochery tinge in the general soil, like what the red sandstone districts on the earth may possibly offer to the inhabitants of Mars." Others suppose it to indicate the existence of a very dense atmosphere, which analyzes the light reflected from the planet.

When the sunlight passes through vapor or clouds in the morning or evening, the different rays of which it is composed are separated, and the *red rays* only pass to the

---

172. Usual appearance of Mars? When brightest, and why? Apparent diameter? Cause of great variation? Phases? Herschel's remark? Spot at north pole?
173. Supposed causes of his *color?* (Note.)

earth, giving to the clouds a gorgeous crimson appearance. In a similar manner it is supposed that the atmosphere of Mars may give him his crimson hue.

TELESCOPIC APPEARANCES OF MARS.

1. The right-hand figure represents Mars as seen at the Cincinnati Observatory, August 5, 1845. On the 30th of tho same month he appeared as represented on the left. The *middle view* is from a drawing by Dr. Dick.

2. Just east of the "Seven Stars," or *Pleiades*, the student will find another group called the *Hyades;* one of which, called *Aldebaran*, is of a reddish cast, and somewhat resembles the planet Mars. When Mars is in opposition, however, at his nearest point to us, and with his enlightened side toward us, he appears much larger and brighter than Aldebaran.

174. As the periodic time of Mars is only 1 yr. 322 days (71), his motion eastward among the stars will be very rapid, as in that time he must traverse the whole circle of the heavens. His rate of motion being about 1° for every two days, or one whole sign in 57 days, it will be easy to detect his eastward progress by observing his change of position with reference to the fixed stars, for a few evenings only; and by marking his place occasionally for two years, we may track him quite around the heavens.

THE ASTEROIDS.

175. The *Asteroids* are invisible except through telescopes, though Vesta was once seen by Schröeter with the naked eye. Few of them present any sensible disks, even under the telescope. They have a pale ash-color, with the exception of Ceres, which is of a reddish hue, resembling Mars. A thin haze or nebulous envelope has been observed around Pallas, supposed to indicate an extensive atmosphere; but no spots or other phenomena have ever been detected.

"On such planets," says Sir John Herschel, "giants might exist; and those animals which on earth require the buoyant power of water to counteract their weight might

---

174. What said of the *eastward motion* of Mars? How detected? Rate?
175. Are the asteroids visible to naked eye? Schröeter? How appear under telescope? Ceres? Pallas? (Remarks of Sir John Herschel?)

there be denizens of the land. A man placed on one of these planets might spring with ease to the hight of 60 feet, and sustain no greater shock in his descent than he does on the earth from leaping a yard." See 65 to 67, and notes.

JUPITER.

176. To the naked eye, Jupiter appears like a fine bright star of the first magnitude. His apparent diameter varies from 30" to 46", according to his distance from the earth. His color is of a pale yellow. Under a telescope, his oblateness is plainly perceptible (as shown at 135), and his disk is seen to be streaked with curious *belts*, running parallel to his equator, as shown in the cut.

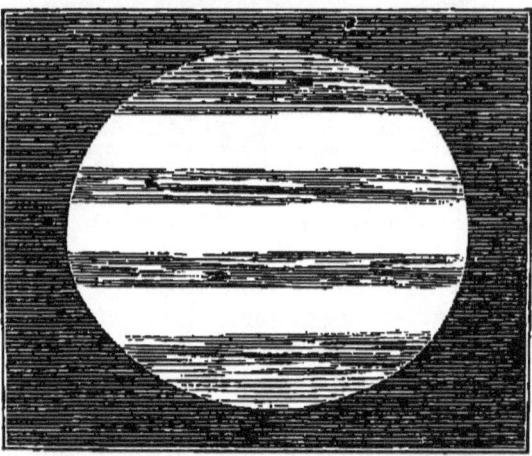

TELESCOPIC VIEW OF JUPITER.

1. The number of belts to be seen upon the disk of Jupiter depends very much upon the power of the instrument through which he is viewed. An ordinary telescope will show the two main belts, one each side of his equator; but those of greater power exhibit more of these curious appendages. Dr. Herschel once saw his whole disk covered with small belts.

177. These belts sometimes continue without change for months, and at other times break up and change their forms in a few hours. They are quite *irregular*, both in *form* and *apparent density;* as both bright and dark spots appear in them, and their edges are always broken and uneven. They are supposed by some to be openings in the atmosphere of the planet, through which its real body is seen; while others think they may be *clouds*, thrown into parallel strata by the rapid motion of Jupiter upon his axis. The *spots* in the belts are thought to

176. Jupiter to naked eye? Apparent magnitude? Cause of variation? Color? Figure? Belts? (Number of belts? Ordinarily? As seen by Herschel? What view in the cut?)

177. Are these belts *permanent* and *regular?* What supposed to be? What said of *spots* in the belts? What ascertained by observing spots? (What said in note?)

be *caverns* or *mountains*, or, at least, something permanently attached to the body of the planet. It was by watching these that the rotation of the planet upon his axis was ascertained.

<small>One of these spots, first observed in 1665, disappeared, and reappeared regularly in the same form for more than forty years; showing conclusively that it was something permanent, and not a mere atmospherical phenomenon.</small>

178. In examining Jupiter with a telescope, from one to four small stars will be seen near him, which, on examination, will be found to accompany him in his eastward journey around the heavens, and to revolve statedly around him. These are the moons of Jupiter, of which we shall speak more fully under the head of Secondary Planets.

<small>The writer once saw all four of these satellites at once, and very distinctly, through a common ship telescope, worth only twelve or fifteen dollars. They were first seen by Galileo with a telescope, the object-glass of which was only one inch in diameter! If the student can get hold of any such instrument whatever, let him try it upon Jupiter, and see if he cannot see from one to four small stars near him, that will occupy different positions at different times.</small>

179. As the periodic time of Jupiter is 11 years 317 days (71), his rate of motion eastward through the fixed stars is about 30° a year. Still, this motion can soon be detected, and in 12 years we may watch his progress quite around the heavens.

<small>The writer has watched this planet from the constellation Aries, west of the "Seven Stars," till he passed that group, and onward through ♉, ♊, ♋, &c., to ♍, his present position (1853). In five years (1858) he will get around to Aries again, where he was seen in 1846; and thenceforward will perform the same journey again every twelve years.</small>

SATURN.

180. This planet is plainly visible to the naked eye, appearing like a star of the third magnitude, of a pale bluish tint. His average angular diameter is about 18″. By the aid of the telescope, he is found not only to be oblate, and striped with belts, and attended by satellites like Jupiter, but to be encircled by a suite of gorgeous *rings*, which renders him one of the most interesting objects in all the heavens.

<small>178. What else discovered about Jupiter? What are they? (Remark in note?)
179. Jupiter's rate of motion eastward? Is it easily detected? (Remark in note? Where was Jupiter in 1846? In 1853? Where *now?* Where in 1858?)
180. Natural appearance of Saturn? Angular diameter? Appearance through telescope?</small>

181. The oblateness of Saturn (15) is distinctly visible through good telescopes (as shown in the cut), while the body of the planet is of a *lead color*, and the rings of a *silvery white*. They may be compared to concentric circles (18) cut out of a sheet of tin. They are broad, flat, and thin, and are placed one within the other directly over the equator of the planet, and revolve with him about his axis, in the same direction, and in the same time (128). They are estimated to be about 100 miles in thickness.

182. These rings are *solid matter*, like the body of the planet. This is proved by the fact that they invariably cast a strong shadow themselves upon the body of the planet, and frequently exhibit the planet's shadow very distinctly upon their own surfaces. It is also evident that they are *wholly detached from the planet*, as the fixed stars in the distant heavens beyond have been seen through the opening in the rings, and between the planet and the first ring.

TELESCOPIC VIEW OF SATURN.

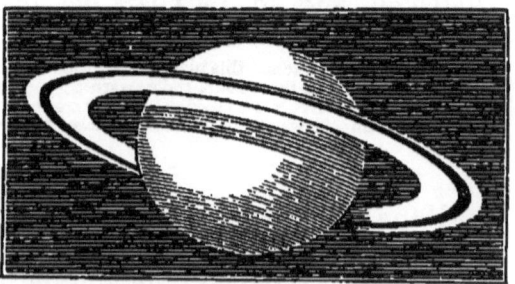

The adjoining cut is an excellent representation of Saturn as seen through a telescope. The oblateness of the planet is easily perceptible, and his *shadow* can be seen upon the rings back of the planet. The shadow of the rings may also be seen running across his disk. The writer has often seen the opening between the body of the planet and the interior ring as distinctly as it appears to the student in the cut. Under very powerful telescopes, these rings are found to be again subdivided into an indefinite number of concentric circles, one within the other, though this is considered doubtful by Sir John Herschel.

183. As our view of the rings of Saturn is generally an oblique one, they usually appear *elliptical*, and never circular. The ellipse seems to contract for about $7\frac{1}{2}$ years, till it almost entirely disappears, when it begins

---

181. Oblateness? Color? Rings—what like? How situated? What motion? Thickness?
182. What said of *substance* of the rings? What proof? What evidence that they are detached? (Remark of author as to seeing satellites? Respecting rings? Opinion of Herschel?)
183. What the general apparent figure of the rings? Why elliptical?

to expand again, and continues to enlarge for 7½ years, when it reaches its maximum of expansion, and again begins to contract. For fifteen years, the part of the rings *toward* us seems to be *thrown up*, while for the next fifteen it appears to drop *below* the apparent center of the planet; and while shifting from one extreme to the other, the rings become almost invisible, appearing only as a faint line of light running from the planet in opposite directions. The rings vary also in their inclination, sometimes dipping to the right, and at others to the left.

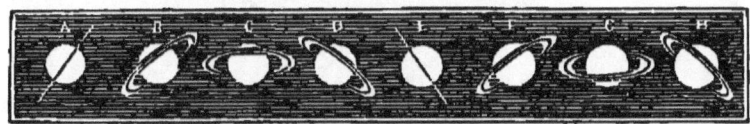

TELESCOPIC PHASES OF THE RINGS OF SATURN.

The above is a good representation of the various inclinations and degrees of expansion of the rings of Saturn, during his periodic journey of 80 years.

184. The rings of the planet are always directed more or less toward the earth, and sometimes exactly toward us; so that we never see them perpendicularly, but always either exactly edgewise, or obliquely, as shown in the last figure. Were either pole of the planet exactly toward us, we should then have a perpendicular view of the rings, as shown in the adjoining cut.

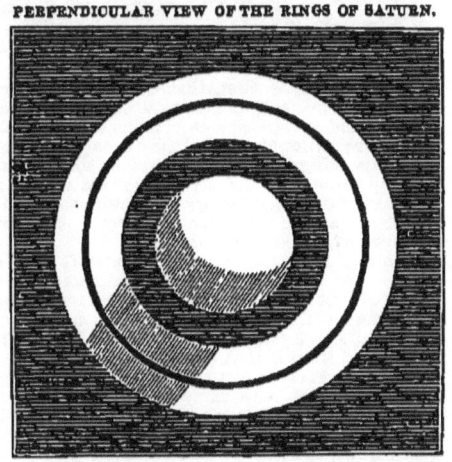

PERPENDICULAR VIEW OF THE RINGS OF SATURN.

185. The various phases of Saturn's rings are explained by the facts that his axis remains parallel to itself (see following cut), with a uniform inclination to the

---

What periodic variation of expansion? Of inclination? When nearly invisible?
184. How are the rings situated with respect to the earth? How would they appear if either pole of Saturn were toward us?

plane of his orbit (122), which is very near the ecliptic (108); and as the rings revolve over his equator, and at right angles with his axis, they also remain parallel to themselves. The revolution of the planet about the earth every 30 years (72) must therefore bring first one side of the rings to view, and then the other—causing all the variations of expansion, position, and inclination which the rings present.

SATURN AT DIFFERENT POINTS IN HIS ORBIT.

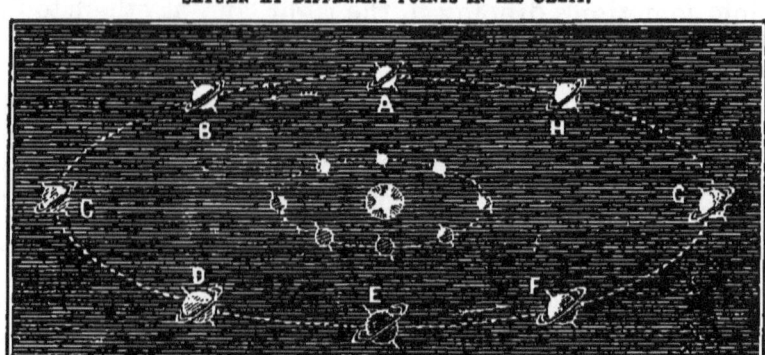

1. Here observe, first, that the axis of Saturn, like those of all the other planets, remains permanent, or *parallel with itself;* and as the rings are in the plane of his equator, and at right angles with his axis, they also must remain parallel to themselves, whatever position the planet may occupy in its orbit.

2. This being the case, it is obvious that while the planet is passing from A to E, the sun will shine upon the *under* or *south* side of the rings; and while he passes from E to A again, upon the *upper* or *north* side; and as it requires about 30 years for the planet to traverse these two semicircles, it is plain that the alternate day and night on the rings will be 15 years each.

3. A and E are the *equinoctial,* and C and G the *solstitial* points in the orbit of Saturn. At A and E the rings are *edgewise* toward the sun, and also toward the earth, provided Saturn is in opposition to the sun. To an observer on the earth, the rings will seem to expand from A to C, and to contract from C to E. So, also, from E to G, and from G to A. Again: from A to E the front of the rings will appear *above* the planet's center, and from E to A *below* it.

4. The rings of Saturn were invisible, as rings, from the 22d of April, 1848, to the 19th of January, 1849. He came to his equinox September 7, 1848; from which time to February, 1856, his rings will continue to expand. From that time to June, 1863, they will contract, when he will reach his other equinox at E, and the rings will be invisible. From June, 1863, to September, 1870, they will again expand; and from September, 1870, to March, 1877, they will contract, when he will be at the equinox passed September 7, 1848, or 29½ years before.

5. The writer has often seen the rings of Saturn in different stages of expansion and contraction, and *once* when they were almost directly *edgewise* toward the earth. At that time (January, 1849), they appeared as a bright line of light, as represented at A and E, in the above cut.

185. What is the *cause* of these varying phases, &c.? (Explain by diagram. When rings invisible? When at his equinox? How long rings expand? Contract? When rings next invisible? Expansion again? Contraction? At what point then? Author's observations?)

SATURN.

186. The dimensions of the rings of Saturn may be stated in round numbers as follows:

	Miles.
Distance from the body of the planet to the first ring	19,000
Width of interior ring	17,000
Space between the interior and exterior rings	2,000
Width of exterior ring	10,500
Thickness of the rings	100

These statistics, as given by Sir John Herschel, are as follows:

Exterior diameter of exterior ring	40″·095	= 176,418	miles.
Interior do.	35″·289	= 155,272	"
Exterior diameter of interior ring	34″·475	= 151,690	"
Interior do.	26″·668	= 117,339	"
Equatorial diameter of the body	17″·991	= 79,160	"
Interval between the planet and interior ring	4″·339	= 19,090	"
Interval of the rings	0″·408	= 1,791	"
Thickness of the rings not exceeding		250	"

187. The rings of Saturn serve as reflectors to reflect the light of the sun upon his disk, as our moon reflects the light to the earth. In his nocturnal sky, they must appear like two gorgeous arches of light, bright as the full moon, and spanning the whole heavens like a stupendous rainbow.

NIGHT SCENE UPON SATURN.

In the annexed cut, the beholder is supposed to be situated some 30° north of the equator of Saturn, and looking directly south. The *shadow* of the planet is seen travelling up the arch as the night advances, while a *New Moon* is shown in the west, and a *Full Moon* in the east at the same time.

188. The two rings united are nearly 13 times as wide as the diameter of the moon; and the nearest is only $\frac{1}{12}$th as far from the planet as the moon is from us.

1. The two rings united are 27,500 miles wide; which ÷ 2,160 the moon's diameter = $12\frac{7}{16}$. So 240,000 miles, the moon's distance ÷ 19,000 the distance of Saturn's interior ring = $12\frac{12}{19}$.

2. At the distance of only 19,000 miles, our moon would appear some forty times as large as she does at her present distance. How magnificent and inconceivably grand, then, must these vast rings appear, with a thousand times the moon's magnitude, and only one-twelfth part of her distance!

186. State the distances and dimensions of his rings, beginning at the body of the planet, and passing outward. (What additional statistics from Herschel?)
187. What purpose do the rings of Saturn serve? How appear in his evening sky?
188. Width of two rings, as compared with moon? Distance? (Demonstrate both. How would our moon appear at the distance of Saturn's rings?)

189. Besides the magnificent rings already described, the telescope reveals *eight satellites* or *moons*, revolving around Saturn. But these are seen only with good instruments, and under favorable circumstances.

On one occasion, the writer saw five of them at once, with a six-inch refractor manufactured by Mr. Henry Fitz, of New York; but the remaining three he has never seen. For a further description of these satellites, see chapters on the Secondary Planets.

190. The periodic time of Saturn being nearly 30 years (72), his motion eastward among the stars must be very slow, amounting to only 12° a year, or one sign in $2\frac{1}{2}$ years. It will be easy, therefore, having once ascertained his position, to watch his slow progress eastward year after year. Saturn is now (October, 1852) about 15° west of the seven stars, and consequently will pass them eastward early in 1854.

URANUS.

191. Uranus is scarcely ever visible except through a telescope; and even then we see nothing but a small round uniformly illuminated disk, without rings, belts, or discernible spots. His apparent diameter is about 4″, from which he never varies much, owing to the smallness of our orbit in comparison with his own.

Sir John Herschel says he is without discernible spots, and yet in his tables he lays down the time of the planet's rotation (which could only be ascertained by the rotation of spots upon the planet's disk), at $9\frac{1}{4}$ hours (128). This time is probably given on the authority of Schröeter, and is marked as doubtful by Dr. Herschel.

192. The motion of Uranus in longitude is still slower than that of Saturn. His periodic time being 84 years 27 days, his eastward motion can amount to only about 4° 17′ in a whole year. To detect this motion requires instruments and close observations. At this date (1853) Uranus has passed over about $\frac{7}{8}$ of his orbit, since his discovery in 1781; and in 1865 will have traversed the whole circuit of the heavens, and reached the point where Herschel found him 84 years before.

---

189. What else seen about Saturn? When seen? (Observations of the author.)

190. Motion of Saturn eastward? Rate?

191. How Uranus seen? How appear through telescopes? Apparent diameter? Why so small, when so much larger than Venus? Why so little variation? (Remark respecting spots.)

192. What said of Uranus' apparent *motion?* Rate per year? In 1853, how far since discovered? When made a complete revolution since 1781?

193. Uranus is attended by several satellites—four at least, probably five or six.

Sir William Herschel reckoned six, though no other observer has confirmed this opinion; and even his son, Sir John Herschel, seems to consider the existence of six satellites quite doubtful.

NEPTUNE.

194. Neptune is a purely telescopic planet, and his immense distance seems to preclude all hope of our coming at much knowledge of his physical state. A single satellite has been discovered in attendance upon him, and the existence of another is suspected; but if others exist, they are as yet undetected.

195. On the 3d of October, 1846, Mr. Lassell, of Liverpool, England, supposed he had discovered a *ring* about the planet, similar to the rings of Saturn; but this supposition has not yet been confirmed by the observations of other astronomers.

196. The periodic time of Neptune being 164 years 226 days, his motion in longitude amounts to only about 2° 10′ per year; and yet this slow motion of about 21″ per day is easily detected, in a short time, by the aid of the proper instruments. It is by this motion, as well as by the disk which it exhibits under the telescope, that the object was first distinguished from the fixed stars, and recognized as a planet.

THE SOLAR SYSTEM IN MINIATURE.

197. Choose any level field or bowling-green, and in its center place a globe two feet in diameter, to represent the sun. Mercury may then be represented by a *mustard-seed*, at the distance of 82 feet; Venus by a *pea*, at the distance of 142 feet; the earth also by a pea, at the distance of 215 feet. A large *pin's head* would represent Mars, if placed 327 feet distant; and the Asteroids may be represented by *grains of sand*, from 500 to 600

---

193. Attendants of Uranus? How many? (Remark in note?)
194. How Neptune seen? What attendant? Suspicion?
195. Supposition of Lassell? Is it confirmed?
196. Motion of Neptune per year? Why so slow? Can it be detected?
197. What representation of the solar system? Size of sun? Mercury.

feet from the center. A moderate sized *orange* would represent Jupiter, at the distance of 80 rods, or 1,320 feet; while a smaller orange would represent Saturn, at the distance of 124 rods, or 2,046 feet. Place a full-sized *cherry* or small *plum* three-fourths of a mile distant for Uranus, and another a mile and a quarter distant for Neptune, and you have the solar system in miniature.

198. To imitate the *motions* of the planets in their orbits, in the above illustration, Mercury must move to the amount of *his own diameter* in 41 seconds; Venus, in 4m. 14s.; the earth, in 7m.; Mars, in 4m. 48s.; Jupiter, in 2h. 56m.; Saturn, in 3h. 13m.; Uranus, in 2h. 16m.; and Neptune, in 3h. 30m.

CHAPTER IV.

SEASONS OF THE DIFFERENT PLANETS, ETC.

199. The general philosophy of the *seasons* has already been explained (Art. 119 to 125).

The *inclination of the axis* of a planet determines the extent and character of its *zones;* and the length of its *periodic time* determines the *length* of its *seasons.*

Thus the axis of the earth being inclined toward the ecliptic 23° 28', the tropics fall 23° 28' from the equator, and the polar circles 23° 28' from the poles; and the period of the earth's revolution around the sun being 365¼ days, it follows that each of the four seasons must include about three months, or 91 days on an average. If the axis was more inclined, the tropics would fall farther from the equator, and the polar circles further from the poles, so that the torrid and frigid zones would be wider, and the temperate narrower; and if the earth's period was longer, her seasons, respectively, would be longer.

200. The *general temperature* of a planet is probably governed by its distance from the sun (59, 60); but the temperature of any particular portion of a planet depends mainly upon *the directness or obliquity with which*

and where placed? Venus, what and where? Earth? Asteroids? Mars &c.

198. How imitate the *motions* of the several planets?
199. What determines the extent and character of a planet's *zones?* What the *length* of its seasons? (Illustrate by inclination and period of the earth.)

the rays of light fall upon it—a circumstance that greatly affects the *amount* of light received by any given portion of its surface. Hence we have summer in the northern hemisphere in July, when the earth is farthest from the sun; and winter in January, when she is nearest the sun (144).

Though nearer the sun in January than in July, still, as the northern hemisphere is then inclined *from* the sun, his rays strike its surface obliquely; less light falls upon the same space than if its contact was more direct, and it is consequently cold. But in July, the rays are more direct—the northern hemisphere being inclined *toward* the sun—and it is summer, notwithstanding we are three millions of miles further from the sun than in January.

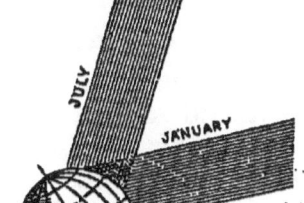

SUMMER AND WINTER RAYS.

1. The comparative amount of light received in the northern hemisphere in July and January may be illustrated by the accompanying figure, in which the rays of light at different seasons are represented to the eye. In January, they are seen to strike the northern hemisphere obliquely, and consequently the same amount of light is spread over a much greater surface. In July, the rays fall almost perpendicularly upon us, and are much more intense. Hence the variations of temperature which constitute the seasons.

2. If the student is not perfectly clear as to *how* the north pole is turned first *toward* and then *from* the sun, he will need to be guarded against the vulgar idea that the earth's axis "wabbles," as it is called. By consulting 119 to 121, and the cuts, it will be seen that the very permanency of a planet's axis, combined with its periodic revolution, gives the beautiful and ever welcome changes of the seasons. How simple, and yet how effectual, this Divine mechanism!

201. As the inclination of the axis of a planet and the length of its periodic time determine the extent and character of its zones, and the length of its seasons, it follows that where these are known, we have a reliable clew to the seasons of a planet, even though we have neither visited nor heard from it; and as we do not know the inclination of the axis of Mercury, we have no knowledge of his seasons.

---

200. What governs the general temperature of the planets? The temperature of particular zones? What result from this last? Why not warmest in January, &c.? (Illustrate by diagram.) How are the poles shifted *to* and *from* the sun? Do the poles "wabble?"

201. How ascertain the character of the seasons of distant planets? Seasons of Mercury?

202. The seasons of *Venus* are very remarkable. So great is her inclination (122), that her tropics fall within 15° of her poles, and her polar circles (as if to retaliate for the trespass upon their territory), go up to within 15° of her equator. Thus the torrid and frigid zones overlap each other, and the temperate zone is altogether annihilated.

The period of Venus being but 225 days (72), the sun declines in that time from her equinoctial to within 15° of one pole; then back to the equinoctial, and to within 15° of the other pole, and again back to the equinoctial. The *effect* of this very great inclination is to give *eight seasons* at her equator every 225 days.

<small>In her short period of 225 days, the sun seems to pass from her northern solstice through her equinox to her southern solstice, and back to the point from which he started. When he is over one of her tropics, it is winter not only at the other tropic, but also at her equator; and as the sun passes over from tropic to tropic, and back again every 225 days, making spring at the equator as he approaches it, summer as he passes over it, autumn as he declines from it, and winter when he reaches the tropic, it follows that at her equator Venus has *eight seasons* in one of her years, or in 225 of our days. Her seasons, therefore, at her equator, consist of only about four weeks of our time, or 28¼ days; and, from the heat of summer to the cold of winter, can be only about 56 days. At her tropics, she has only four seasons of 56 days each.</small>

203. The polar inclination of *Mars* being 28° 40′ (122), his torrid zone must be 57° 40′ from his poles—leaving only 32° 40′ for the width of his temperate zone. But as his year consists of 687 days, his four seasons must consist of about 172 days each, or nearly twice the length of the seasons of our globe.

204. So slight is the inclination of the axis of *Jupiter* to his orbit, that he has but a narrow torrid zone, and small polar circles. As his orbit departs from the plane of the ecliptic only 1° 46′ (108), and his axis is inclined to his orbit only 5° 3′, it follows that his axis is nearly perpendicular to the ecliptic. The sun never departs more than 5° 3′ from his equator; and still, as his periodic time is about 12 years (72), he has alternately six years of northern and six of southern declination. His narrow torrid zone and small polar circles leave very ex-

<small>202. Seasons of Venus? Where her tropics? Polar circles? Temperate zone? Sun's declination upon her? Its effect? (Substance of note?)
203. Zones of Mars? Length of seasons, and why?
204. Zones of Jupiter, and why? Describe his climate. Seasons? Days and nights? Poles?</small>

tensive temperate zones. In passing from his equator to his poles, we meet every variety of climate, from the warmest to the coldest, with but slight variations in any latitude, from age to age. His days and nights are always nearly of the same length, as the sun is always near his equinoctial. His poles have alternately six years day and six years night.

205. The polar inclination and zones of *Saturn* differ but little from those of Mars; but his seasons are greatly modified by the length of his periodic time. This being about 30 years, his four seasons must each be about $7\frac{1}{2}$ years long; and his polar regions must have alternately 15 years day and 15 years night. The rings of Saturn, which lie in the plane of his equator, and revolve every $10\frac{1}{4}$ hours, are crossed by the sun when he crosses the equinoctial of the planet. During the southern declination of the sun, which lasts fifteen years, the south side of the rings is enlightened, and has its summer. It has also its day and night, by revolving in a portion of the planet's shadow. When the sun is at the southern tropic, it is midsummer on the south side of the rings, as the rays of light then fall most directly upon them. As the sun approaches the equator, the temperature decreases, till he crosses the equinoctial, and the long winter of fifteen years begins. At the same time, the north side of the rings begins to have its spring; summer ensues, and in turn it has fifteen years of light and heat. The influence of these wonderful rings upon the climate of Saturn must be very considerable. During the winter in each hemisphere, they cast a deep shadow upon some portion of his surface during the day; and in the summer, these immense reflectors so near the planet, and so bright in the sunlight, must contribute greatly to the light, if not to the warmth, of his summer evenings. The poles of Saturn are alternately 15 years in the light, and 15 years in darkness.

206. Of the inclination of the axes of Uranus and

---

205. Zones of Saturn, and why? Length of seasons? Rings—how enlightened? Influence upon climate? Polar days and nights?

Neptune, respectively, we have no knowledge, and consequently can form no opinion respecting their tropics, polar circles, zones, &c. If not too much inclined, like Venus, they have but four seasons in *their year*, which would make each season of Uranus 22 years and 9 days long, and each season of Neptune 41 years and 56½ days long; as these periods are, respectively, one-fourth of the periodic time of the planet (72).

<small>Thus we see that tropics, polar circles, zones, and seasons are not peculiar to our globe, but are a necessary result of an inclined axis, and a revolution around the sun. The *causes* which produce our seasons are known to be in operation in other planetary worlds, and it would be unreasonable to deny that the *effect* was there also.</small>

DISCOVERY OF THE DIFFERENT PLANETS.

207. The old planets, as they are called, viz., *Mercury, Venus, Mars, Jupiter,* and *Saturn,* have been known as planets, or "wanderers," from the earliest ages. *Uranus* was discovered by Sir William Herschel, March 13th, 1781. *Neptune was demonstrated to exist before it had been seen,* by M. Le Verrier, of France, August, 1846; and first seen by Dr. Galle, of Berlin, Sept. 23, 1846.

208. The discovery of Neptune is probably one of the greatest achievements of mathematical science ever recorded. By comparing the true places of Uranus with the places assigned by the tables, it was found that he was not where his known rate of motion required him to be; and after making all due allowance for the attraction of Jupiter and Saturn (65), by which perturbations would be produced, it was found that there was evidently the *effect* of some other body, *exterior* to the orbit of Uranus, the attraction of which body helped to cause the perturbations of Uranus. From this *effect,* produced by an unknown and invisible world, lying far out beyond the supposed boundaries of the solar system, not only was the *existence* of its cause demonstrated, but its *direction, distance, mass,* and *period* were proximately ascertained.

<small>206. What said of the seasons of Uranus and Neptune? Probable length of former? Latter? (Remark in note?)
207. What said of the "old planets?" Of Uranus? Neptune?
208. Describe the discovery of Neptune. Perturbations? Tables, &c.? (Describe successive steps in detail. What said of Mr. Adams?)</small>

DISCOVERY OF THE DIFFERENT PLANETS. 101

1. On the evening of the 23d of September, 1846, Dr. Galle, one of the astronomers of the Royal Observatory at Berlin, received a letter from Le Verrier, of Paris, requesting him to employ the great telescope at his command in searching for the supposed new planet, and giving its position, as ascertained by calculation, as 325° 52·8' of geocentric longitude. Dr. Galle, taking advantage of the very evening on which he received Le Verrier's letter, soon discovered an object resembling a star of the eighth magnitude, near the spot indicated by Le Verrier, as the place of the new planet. On consulting an accurate star chart, it was found that no such star was there laid down, and observations were at once commenced, with a view to detecting any change of place. In three hours time, it was seen to have moved; and by the next evening at eight o'clock, it was found to have retrograded more than four seconds of time (see 97 and cut)—a circumstance which proved it to be much nearer the earth than the fixed stars, and consequently a planet—the very planet which had caused the unaccountable irregularities of Uranus. The geocentric longitude of the planet, at midnight, September 23, 1846, was 325° 52·8'; which was less than 1° from the place assigned to it by Le Verrier! The reason why Le Verrier wrote to Dr. Galle was, that the former had no suitable telescope for conducting the search in which he was so deeply interested.

2. It is worthy of remark that Mr. Adams, of St. John's College, Cambridge. England, had also calculated the place, &c., of the new planet, and had arrived at results similar to those reached by Le Verrier; but as the latter had *published* his conclusions first, the honor of the discovery is generally accorded to Le Verrier.

209. The *Asteroids* have all been discovered during the present century, and most of them since 1847. And to the number now known, it is not improbable that others will be added from time to time.

The following table will show the date, &c., of the discovery of the several asteroids. They are laid down in the order of discovery:

Planet.	Date.	Discoverers.
Ceres	January 1, 1801	Piazzi, of Palermo
Pallas	March 28, 1802	Olbers, of Bremen
Juno	September 1, 1804	Harding, of Bremen.
Vesta	March 29, 1807	Olbers, of Bremen.
Astræa	December 8, 1845	Encke, of Dresden.
Hebe	July 5, 1847	" "
Iris	August 13, 1847	Hind, of London.
Flora	October 18, 1847	" "
Metis	April 25, 1848	Graham, of Sligo, Ireland.
Hygeia	" "	Gasparis, of Naples, Italy.
Parthenope	May 11, 1850	" "
Clio	September 13, 1850	Hind, of London.
Egeria	November 2, 1850	Gasparis, of Naples.
Irene	May 19, 1851	Hind, of London.
Eunomia	July 29, 1851	Gasparis, of Naples.
Melpomene	June 24, 1852	Hind, of London.
Anonymous*	August 22, 1852	" "

* At the date of this writing, it is only *forty days* since this last planet was discovered; hence it has no name as yet. The same fact accounts for the absence of the last two of this list from most of the preceding tables.

---

209. What said of the discovery of the asteroids? Are there probably others?

CHAPTER V.

SECONDARY PLANETS—THE MOON.

210. The *Secondary Planets* are those that revolve statedly around the primaries, and accompany them in their periodical journeys around the sun. Of these, the earth has one; Jupiter, four; Saturn, eight; Uranus, six; and Neptune, one—in all twenty. Besides these, there is a strong suspicion among astronomers that Venus is attended by a satellite, and that Neptune has at least two, instead of one.

Sir John Herschel says Uranus is attended "certainly by four, and perhaps by six, and Neptune by two or more." Outlines, Art. 533. In regard to Venus, Prof. Hind, of London, says: "Astronomers are by no means satisfied whether Venus should be attended by a satellite or not. * * * It is a question of great interest, and must remain open for future discussion."

211. Though the secondary planets have a compound motion, and revolve both around the sun and around their respective primaries, they are subject to the same general laws of gravitation—of centripetal and centrifugal force—by which their primaries are governed. Like them, they receive their light and heat from the sun, and revolve periodically in their orbits, and on their respective axes. In the economy of nature, they seem to serve as so many *mirrors* to reflect the sun's light upon superior worlds, when their sides are turned away from a more direct illumination.

The design of all the secondaries may be inferred from what is said of the purposes for which our own satellite was created. "And God said, Let there be lights in the firmament of heaven, to divide the day from the night; and let them be for signs, and for seasons, and for days and years; and let them be for lights in the firmament of heaven, to give light upon the earth: and it was so. And God made two great lights; the greater light to rule the day, and the lesser light to rule the night: he made the stars also."—Gen. i, 14—16.

---

210. What are the Secondary planets? How many? How distributed? What supposition respecting Venus? Neptune? (Herschel's remark? Prof. Hind's?)

211. What said of the laws by which the primaries are governed? Light and heat? Uses? (From what may we infer their design?)

THE MOON.

212. To the inhabitants of our globe, the earth's satellite or moon is one of the most interesting objects in all the heavens. Her nearness to the earth, and consequent apparent magnitude, her rapid angular motion eastward, her perpetual phases or changes, and the mottled appearance of her surface, even to the naked eye, all conspire to arrest the attention, and to awaken inquiry. Add to this her connection with *Eclipses*, and her influence in the production of *Tides* (of both of which we shall speak hereafter in distinct chapters), and she opens before us one of the most interesting fields of astronomical research.

213. The Romans called the moon *Luna*, and the Greeks *Selene*. From the former, we have our English terms *lunar* and *lunacy*. In mythology, Selene was the daughter of *Helios*, the sun. Our English word *selenography*—a description of the moon's surface—is from *Selene*, her ancient name, and *grapho*, to describe.

214. The point in the moon's orbit nearest the earth is called *Perigee*, from the Greek *peri*, about, and *ge*, the earth. The point most distant is called *Apogee*, from *apo*, from, and *ge*, the earth. These two points are also called the *apsides* of her orbit; and a line joining them, the *line of the apsides*.

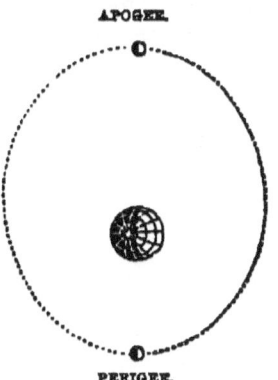

See the moon in apogee and perigee in the cut. The singular of apsides is *apsis*.

215. The mean *distance* of the moon from the earth's center is, in round numbers, 240,000 miles; or, more accurately, 238,650. The eccentricity of her orbit amounting to 13,333 miles, of course her distance must vary, and also her apparent magnitude (56). Her average angular

---

212. What said of our moon? Why specially interesting?
213. Latin name of the moon? Greek? Derivation of words from Luna? Who was *Selene* in mythology? Selenography?
214. Perigee and Apogee? Derivation? What other name for these two points? What is the *line* of the apsides? (Apsis?)
215. Moon's distance? Does it vary? Why? Eccentricity of orbit?

diameter is 31′ 7″, and her real diameter 2,160 miles. She is consequently only $\frac{1}{49}$th part as large as the earth, and $\frac{1}{70000000}$th part as large as the sun.

<small>The masses of globes are proportional to the cubes of their diameters. Then 2,160 × 2,160 × 2,160 = 10,077,696,000, the cube of the moon's diameter; and 7,912 × 7,912 × 7,912 = 495,289,174,428, the cube of the earth's diameter. Divide the latter by the former, and we have 49 and a fraction over, as the number of times the bulk of the moon is contained in the earth. Its mass, as compared with the sun, is ascertained in the same manner.</small>

216. The plane of the moon's orbit is very near that of the ecliptic. It departs from the latter only about $5\frac{1}{8}°$ (5° 8′ 48″).

INCLINATION OF THE MOON'S ORBIT TO THE PLANE OF THE ECLIPTIC.

<small>Let the line A B represent the plane of the earth's orbit, and the line joining the moon at C and D would represent the inclination of the moon's orbit to that of the earth. At C the moon would be *within* the earth's orbit, and at D exterior to it; and it would be *Full Moon* at D, and *New Moon* at C.</small>

217. The line of the apsides of the moon's orbit is not fixed in the ecliptic, but revolves slowly around the ecliptic, from west to east, in the period of about nine years.

MOTION OF THE APSIDES.

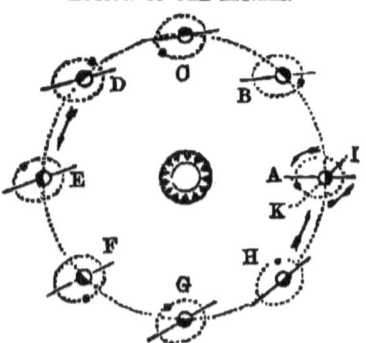

<small>In the adjoining cut, an attempt is made to represent this motion. At A, the line of the apsides points directly to the right and left; but at B, C, and D it is seen changing its direction, till at E the change is very perceptible when compared with A. But the same ratio of change continues; and at the end of a year, when the earth reaches A again, the line of the apsides is found to have revolved eastward to the dotted line I K, or about 40°. In nine years, the aphelion point near A will have made a complete revolution, and returned to its original position.</small>

218. The line of the moon's nodes is also in revolution; but it retrogrades or falls back *westward*, making the circuit of the ecliptic once in about 19 years.

<small>Angular diameter? In miles? How compare with earth? With sun? (How demonstrated?)

216. How is the plane of the moon's orbit situated with respect to the ecliptic? (Illustrate by diagram.)

217. Is the line of the moon's apsides stationary or not? What motion? Period? (Illustrate.)

218. What of the line of the moon's nodes? In what time does it make the circuit of the ecliptic? Amount of motion?</small>

THE MOON. 105

The amount of this motion is 10° 35' per annum, which would require 18 years and 219 days for a complete revolution.

219. The *diameter* of the moon is only $\frac{1}{400}$th part as great as that of the sun; and yet the *apparent* diameter of the moon is nearly equal to that of the sun. The former is 31' 7'', and the latter 32' 2'', or only 55'' difference. The reason why the moon appears to vie with the sun in magnitude, when she is only $\frac{1}{70,000,000}$ as large, is, that she is 400 times nearer to us than he is. See Art. 56.

1. The cut in the margin will show how it is that a small object near us will fill as large an angle, or, in other words, appear as large, as a much larger object which is more remote. The moon at A fills the same angle that is filled by the sun at B.

2. This fact may serve to illustrate the comparative influence of things present and future upon most minds. The little moon may eclipse the sun; or even a dime, if held near enough to the eye, will completely hide all his glories from our view. So in morals and religion. The "things which are seen and temporal" are too apt to hide from our view the more distant but superior glories of the life to come.

220. The *density* of the moon is only about two-thirds that of the earth, and her *surface* $\frac{1}{13}$th as great. The light reflected to the earth by her, at her full, is only $\frac{1}{300,000}$th part as much as we receive on an average from the sun.

221. The daily apparent revolution of the moon is from east to west, with the sun and stars; but her real motion around the earth is from west to east. Hence, when first seen as a "new moon," she is very near, but just east of the sun; but departs further and further from him eastward, till at length she is seen in the east as a full moon, as the sun goes down in the west.

222. The moon performs a *sidereal* revolution around the earth in 27d. 7h. 43m.; and a *synodic* in 29d. 12h. 44m. The sidereal is a complete revolution, as measured by a fixed star; but the motion of the earth eastward in

---

219. Moon's diameter, as compared with that of the sun? With sun's *apparent* diameter? Why appear so near of a size? (Illustrate by diagram. Reflection of the author?)
220. Density of the moon? Her light?
221. Her daily apparent motion? Real motion? How traced?
222. What is her sidereal revolution? Her synodic? What difference? Why? (Illustrate by diagram.)

her orbit gives the sun an apparent motion eastward among the stars (119), and renders it necessary for the moon to perform a little more than a complete revolution each month, in order to come in conjunction with the sun, and make a *synodic* revolution.

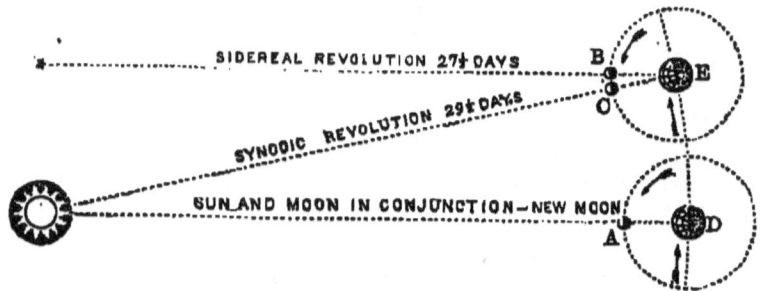

SIDEREAL AND SYNODIC REVOLUTIONS OF THE MOON.

1. On the right, the earth is shown in her orbit, revolving around the sun, and the moon in her orbit, revolving around the earth. At A, the sun and moon are in *conjunction*, or it is *New Moon*. As the earth passes from D to E, the moon passes around from A to B, or the exact point in her orbit where she was 27¼ days before. But she is still west of the sun, and must pass on from B to C, or 1 day and 20 hours longer, before she can again come in conjunction with him. This 1 day and 20 hours constitutes the difference between a sidereal and a synodic revolution.

2. The student will perceive that the difference between a sidereal and synodic revolution of the moon, like that between solar and sidereal time, is due to the same cause—namely, the revolution of the earth around the sun. See 185.

223. The daily angular motion of the moon eastward is 13° 10′ 35″. Her average hourly motion is about 32,300 miles. This motion may be detected by watching her for a few hours only; and by marking her position, with reference to the stars, from night to night, her daily journeys will appear prominent and striking.

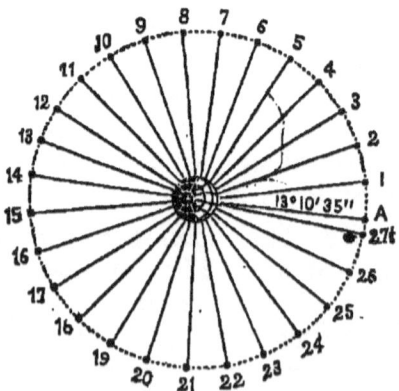

DAILY PROGRESS OF THE MOON EASTWARD.

The estimate of 13° 10′ 35″ is made for a sidereal day of twenty-four hours. In the above cut, the daily progress of the moon may be traced from her conjunction or "change" at A on the right, around to the same point again. This being a *sidereal* revolution, requires only 27¼ days.

223. Daily angular motion eastward? How detected? (For what day is this estimate made?)

224. In her journeyings eastward, the moon often seems to run over and obscure the distant planets and stars. This phenomenon is called an *occultation.*

The adjoining cut represents the new moon as just about to obscure a distant star, by passing between us and it. In 1850, she occulted Jupiter for three revolutions in succession—viz., Jan. 30th, Feb. 27th, and March 26th. Through a telescope, the moon is seen to be constantly obscuring stars that are invisible to the naked eye. They disappear behind the moon's eastern limb, and in a short time reappear from behind her western; thus distinctly exhibiting her eastward motion.

225. Though the moon's orbit is an ellipse, with respect to the earth, it is, in reality, an irregular curve, always *concave toward the sun,* and crossing the earth's orbit every 13° nearly.

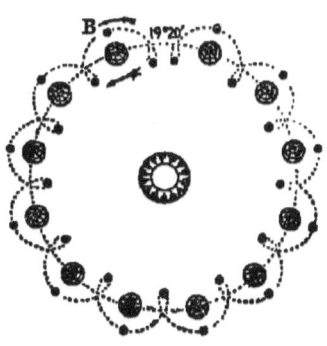

1. If the earth stood still in her orbit, the moon would describe just such a path in the ecliptic as she describes with respect to the earth.
2. If the earth moved but slowly on her way, the moon would actually retrograde on the ecliptic at the time of her change, and would cross her own path at every revolution, as shown in the adjoining figure. But as the earth advances some 46 millions of miles, or near 100 times the diameter of the moon's orbit, during a single lunation, it is evident that the moon's orbit never can return into itself, or retrograde, as here represented.

THE MOON'S ORBIT ALWAYS CONCAVE TOWARD THE SUN.

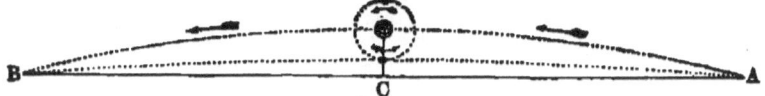

3. That the lunar orbit is always concave toward the sun, may be demonstrated by the above diagram. Let the upper curve line A B represent an arc of the earth's orbit, equal to that passed through by the earth during half a lunation. Now the radius and arc being known, it is found that the chord A B must pass more than 400,000 miles within the earth. But as the moon departs only 240,000 from the earth, as shown in the figure, it follows that she must describe the curve denoted by the middle line, which is concave toward the sun.

---

224. What are *occultations?* How produced? (Are they frequent? Are planets ever occulted? Describe process.)
225. What is the form of the moon's orbit with respect to the earth? The sun? (How if the earth were stationary? If moving slowly? Demonstrate her orbit to be concave, &c. Draw orbit for complete lunation, and describe her relative motion.)

4. This subject may be still further illustrated by the following cut, representing

THE MOON'S PATH DURING A COMPLETE LUNATION.

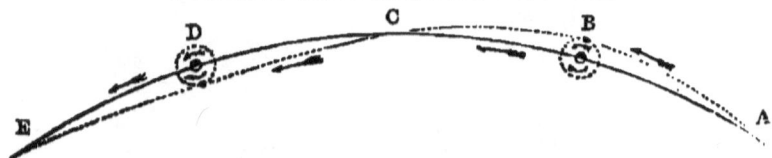

Here the plain line represents the earth's orbit, and the dotted one that of the moon. At A the moon crosses the earth's track 240,000 miles *behind* her. She gains on the earth, till in seven days she passes her at B as a *Full Moon*. Continuing to gain on the earth, she crosses her orbit at C, 240,000 miles *ahead* of her, being then at her *Third Quarter*. From this point the earth gains upon the moon, till seven days afterward she overtakes her at D as a *New Moon*. From D to E the earth continues to gain, till at E the moon crosses 240,000 *behind* the earth, as she had done four weeks before at A. Thus the moon winds her way along, first within and then without the earth; always gaining upon us when outside of our orbit, and falling behind us when within it.

5. The small circles in the cut represent the moon's orbit with respect to the earth, which is as regular *to us* as if the earth had no revolution around the sun.

226. The moon never retrogrades on the ecliptic, or returns into her own path again; but is always advancing with the earth, at the rate of not less than 65,700 miles per hour.

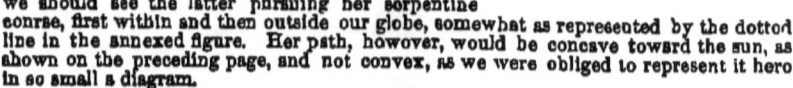

MOON'S PATH.

1. The moon's orbital velocity, with respect to the earth, is about 2,800 miles per hour. When outside the earth, as at B, in the last figure, she *gains* 2,300 miles per hour, which, added to the earth's velocity, would give 70,300 miles as the hourly velocity of the moon. When *within* the earth's orbit, as at D, she *loses* 2,300 miles per hour, which, subtracted from 68,000 miles (the earth's hourly velocity), would leave 65,700 miles as the slowest motion of the moon in space, even when she is falling behind the earth.

2. Could we look down perpendicularly upon the ecliptic, and see the paths of the earth and moon, we should see the latter pursuing her serpentine course, first within and then outside our globe, somewhat as represented by the dotted line in the annexed figure. Her path, however, would be concave toward the sun, as shown on the preceding page, and not convex, as we were obliged to represent it here in so small a diagram.

227. That the moon is opake, like the rest of the planets, and shines only by reflection, is obvious, from the fact that we can see only that part of her upon which the sun shines; and as the enlightened portion is sometimes toward and sometimes from us, the moon is constantly varying in her apparent form and brightness. These variations are called her *phases*.

226. At what rate does the moon advance with the earth? Moon's orbitual velocity, with respect to the earth? Slowest motion? (Illustrate the moon's course.)

227. What proof that the moon is opake? What meant by her *phases?*

228. The *cause* of the moon's phases—her waxing and waning—is her revolution around the earth, which enables us to see more of her enlightened side at one time than at another.

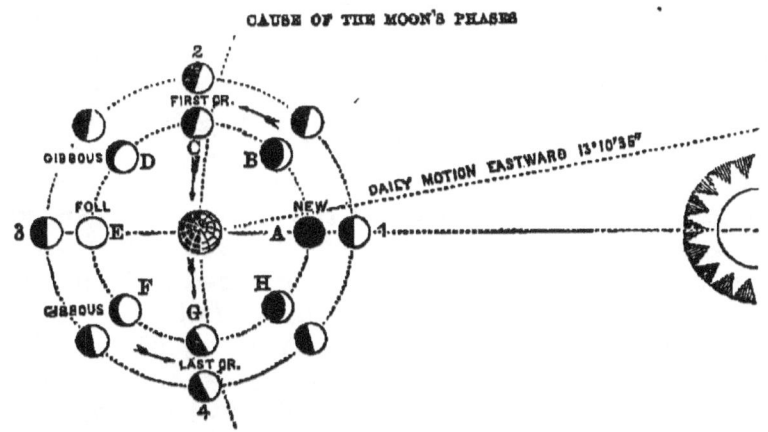

CAUSE OF THE MOON'S PHASES

1. This cut represents the moon revolving eastward around the earth. In the outside circle, she is represented as she would appear, if viewed from a direction at right angles with the plane of her orbit. The side toward the sun is enlightened in every case, and she appears like a half moon at every point.

2. The interior cut represents her as she appears when viewed from the earth. At A it is New Moon; and if seen at all so near the sun, she would appear like a dark globe. At B she would appear like a crescent, concave toward the east. At C, more of her enlightened side is visible; at D, still more; and at E, the enlightened hemisphere is fully in view. We then call her a *Full Moon*. From E around to A again, the dark portion becomes more and more visible, as the luminous part goes out of view, till she comes to her change at A. When at D and F, the moon is said to be *gibbous*.

3. If the student will turn his book bottom upward, and hold it south of him, he will see *why* the crescent of the old moon at H is concave on the west, instead of the east, like the new moon, and why she is seen before sunrise, instead of just after sunset.

229. The *cusps* of the moon are the extremities of the crescent. Her *syzygies* are two points in her orbit 190° apart, where she is new and full moon. (See positions 1 and 3 in the last cut.) The *quadratures* are four points 90° apart (like 1, 2, 3, and 4 in cut); and her *octants* eight points 45° apart (like A, B, C, &c., in the cut).

230. The moon is said to *change* when she comes in conjunction with the sun, and is changed from *Old Moon* to *New Moon*.

---

228. Cause of phases? (Illustrate.)
229. What are the *cusps* of the moon? Her *Syzygies?* *Quadratures?* *Octants?* (Illustrate on blackboard.)
230. What meant by the *change* of the moon? (How noticed or traced?)

If the student will be on the look-out, he can easily find the moon *west* of the sun *in the daytime;* and, by observing her carefully, will see that she is rapidly approaching him. In a short time she will be lost in his beams, and soon after will appear *east* of the sun, just after sundown, as a *New Moon.* This *change,* as it is called, takes place when she passes the sun eastward.

231. A *New Moon* is the moon when she has just passed the sun in her eastward journey, and when only a small portion of her enlightened hemisphere is visible from the earth. She then appears like a slender crescent, concave on the east. The *First Quarter* is when she has advanced 90° eastward from the sun. She is then *south* of us at sundown, and we see one-half of her enlightened side. The *Full* of the moon is when she has advanced 180° from the sun, and is in the *east* when he goes down in the west. Her enlightened side is then toward us, and she appears circular, or full. The *Third Quarter* is when the moon has advanced 270°, or ¾ths of her synodic journey. She has been waning since the full, on her *western* limb, and is now *gibbous.* She is but 90° *west* of the sun, is approaching him, and waning more and more every day. The *waxing* of the moon is from the *change* to the *full;* and the *waning,* from the *full* to the *change* again.

We earnestly recommend to both teacher and student to observe the present place and appearance of the moon, and watch her through one lunation at least. A little time spent in this way will do more to fix correct ideas in the mind than months of abstract study.

232. The line which separates the dark from the enlightened portion of the moon's disk is called the *Terminator.*

As just one-half of the moon is always enlightened by the sun, whether it appears so to us or not, it follows that the terminator must extend quite around the moon, dividing the enlightened from the unenlightened hemisphere. This circle is called the *Circle of Illumination.* At new and full moon this circle is *sidewise* to us; but at the first and third quarters, it is *edgewise.* The portion of the terminator visible from the earth traverses the moon's disk twice during every lunation.

233. A variety of dark lines and spots may be seen upon the surface of the moon with the naked eye. There is a dark figure on her western limb, resembling that of *a man*, with his head to the north, and his body inclined

---

231. What is the *New Moon?* How appear? *First Quarter?* When? Appearance? *Full Moon* and appearance? *Third Quarter?* Position and appearance? When waxing? Waning? (What recommended by author?)
232. What is the *Terminator?* (Substance of note?)
233. Describe the natural appearance of the full moon. (What said of cut? Sketch on blackboard. Ojibway legend?)

to the east. Just east of him, and opposite his shoulders is an irregular object, resembling a huge *bundle* or *pack*.

NATURAL APPEARANCE OF THE FULL MOON.

1. Both these objects are represented in the adjoining cut, which was drawn from nature by the author, on the evening of December 18, 1850. It represents the moon as she appears when about two hours high, and is the best of six different sketches taken during the same evening. Let the student compare it with *the next Full Moon*, and see if our drawing is correct.

2. The Ojibway Indians have a legend by which they explain this singular appearance of the moon. Instead of a "man," they say this figure is a beautiful Ojibway *maiden*, who was translated to the moon "many snows ago," for having set her affections upon that object, and refusing to marry any of the "young braves" of the Ojibway nation. How the "beautiful maiden" came to look so coarse and masculine, and what the rest of the figure means, the tradition does not inform us.

234. These rude figures upon the moon's disk are probably the outlines of her great natural divisions, as mountains, valleys, and continents.

Almost everybody has noticed these rude figures upon the face of the moon, and many, doubtless, have wondered what they were; but how few have supposed, as they were gazing upon her mottled disk, that they were enjoying a distant view of a *world*, and that these dim outlines were a *natural map* of its nearest hemisphere! Having seen the "man in the moon," they have supposed it useless to pursue the subject any further, and here their investigations have ended.

235. By a careful observation of the moon's disk, from month to month, it is found that *the same side is always toward the earth*. From this fact, it follows that she revolves on her axis but once during her synodic revolution around the earth.

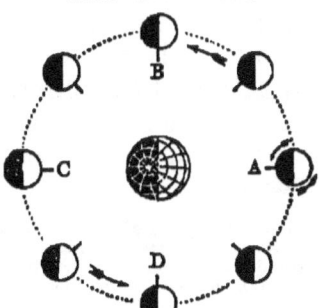

MOON'S REVOLUTION.

1. By watching the moon carefully with the naked eye, it will be seen that the same spots occupy nearly the same places upon her disk from month to month; which shows that the same side is always toward us.

2. Suppose a monument erected upon the moon's surface, so as to point toward the earth at *New Moon*, as represented at A. From the earth it would appear in the moon's center. Now if the moon so revolved upon her axis, in the direction of the arrows, as to keep the pillar pointing directly toward the earth, as shown at A, B, C, and D, and the intermediate points, she must make just one revolution on her axis during her periodic revolution. At A, the pillar points *from* the *sun*, and at C *toward* him; showing that, in going half way round the earth, she has performed half a revolution upon her axis.

---

234. What are these rude figures supposed to be? (Note.)
235. What interesting fact established by watching the moon? What follows from it? (Illustrate by sketch of cut on blackboard.)

236. As the same side of the moon is always toward us, it follows that the earth is invisible from one-half of the moon. From the other half, our globe would appear like a stationary planet, nearly thirteen times as large as the moon appears to us, and exhibiting all her varying phases.

237. Though the moon always presents nearly the same hemisphere toward the earth, it is not always *precisely* the same. Owing to the ellipticity of her orbit, and the consequent inequality of her angular velocity, she appears to *roll* a little on her axis, first one way and then the other—thus alternately revealing and hiding new territory, as it were, on her eastern and western limbs. This rolling motion east and west is called her *libration in longitude*.

MOON'S LIBRATIONS.

The accompanying cut will illustrate the subject of the moon's librations in longitude.
1. From A around to C, the angular motion is *slower* than the average, and the diurnal motion gains upon it, so that the pillar points *west* of the earth, and we see more of the *eastern* limb of the moon.
2. From C to A, again, the moon advances *faster* than a mean rate, and *gains* upon the diurnal revolution; so that the pillar points *east* of the earth, and we see more of the moon's *western* limb. Thus she seems to librate or roll, first one way and then the other, during every periodic revolution.
At B, we see most of her eastern limb; and at D, most of her western.

238. The axis of the moon is inclined to the plane of her orbit only about one and a half degrees (1° 30′ 10·8″). But this slight inclination enables us to see first one pole and then the other, in her revolution around the earth. These slight rolling motions are called her *librations in latitude*.

As the inclination of the earth's axis brings first one pole and then the other toward the *sun*, and produces the seasons, so the inclination of the moon's axis brings first one pole and then the other in view from the earth. But as her inclination is only 1½°, the libration in latitude is very slight.

---

236. What other fact follows from the moon's keeping the same side toward us? How would our globe appear from the moon?
237. What are the moon's *librations?* In *longitude*, and cause? (Illustrate on blackboard.)
238. In *latitude?* Cause? (Illustrate by the case of the earth.)

239. The moon's year consists of $29\frac{1}{2}$ of our days; but as she makes but one revolution upon her axis in that time, she can have but one day and one night in her whole year. And so slight is the inclination of her axis to the plane of her orbit, that the sun's declination from her equator is only about $1\frac{1}{2}°$. She must therefore have perpetual winter at her poles; while at her equator, her long days are very warm, and her long nights very cold.

240. By the aid of the telescope, the surface of the moon is found to be exceedingly rough and uneven, covered with vast plains, deep valleys, and lofty mountains. Several of the latter are from three to four and a half miles high. That they are really mountains is proved by three facts: 1st, the line of the terminator is jagged or uneven, as shown in the cut; 2d., shadows are seen projecting first to the east and then to the west, showing the existence of elevations of some sort, that intercept the light; and 3d, from new to full moon, bright spots break out from time to time, just *east* of the terminator, in the dark portion, and grow larger and larger, till they join the illuminated portion, showing them to be the tops of mountains, which reflect the sunlight before it reaches the intervening valleys.

TELESCOPIC VIEW OF THE MOON.

1. Specimens of these *shadows* may be seen in the cut, projecting to the left. Bright points of light, or, in other words, the illuminated tops of mountains, may also be seen near the terminator, in the dark portion. The writer has often watched them, and seen them enlarge more and more, as the sun arose upon the side of the moon toward us, and enlightened the sides of her mountains.

239. Length of moon's year? Number of natural days? Sun's declination upon her? Climate at equator and poles?
240. How appear through telescopes? What proof of mountains? (Remarks upon cut? Observations of the author? Describe shadows and their changes. Illustrate, by reference to the Andes and their shadows.)

2. The shadows are always projected in a direction opposite the sun, or toward the dark side of the moon; and as her eastern limb is dark from the change to the full, and her western from the full to the change, of course the direction of the shadows must be reversed.

3. Suppose a person stationed at a distance directly over the Andes. Before the sun arose, he would see the tallest peaks enlightened; and as he arose, the long shadows of the mountains would extend to the *west*. At noon, however, little or no shadow would be visible; but at sunset, they would again be seen stretching away to the *east*. This is precisely the change that is seen to take place with the lunar shadows, except that the *time* required is a lunar day, equal to about 15 of our days, instead of one of our days of 12 hours.

241. Some of the lunar mountains are in extensive ranges, like our Alps and Andes; while others are circular, like the craters of huge volcanoes. Great numbers of the latter may be seen with telescopes of only moderate power. Through such an instrument, the moon will appear of a yellowish hue, and the circular mountains like drops of thick oil on the surface of water. Two extensive ranges, and several of the circular elevations, are shown in the last cut. Dr. Scoresby, of Bradford, England, who examined the moon through the monster telescope of Lord Rosse, says he saw a vast number of extinct volcanoes, some of whose craters were several miles in breadth. Her general appearance was that of a vast ruin of nature. Dr. Herschel supposed he saw the light of several *active* volcanoes upon her surface.

242. In regard to the existence of an *atmosphere* around the moon, astronomers are divided. From observations during eclipses of the sun, and other phenomena, it is thought that if the moon has any atmosphere at all, it must be very limited in extent, and far less dense than that of the earth. Dr. Scoresby saw no indications of the existence of water, or of an atmosphere.

From observations during several occultations of stars, the writer is of opinion that a refracting medium of some sort exists in the vicinity of the moon. The *atmosphere* is doubtless subject to the general law of gravitation. Hence it is most dense at the earth's surface, and grows rare as we ascend. Inasmuch, therefore, as the general density of the atmosphere of any planet is dependent upon the attracting force of that planet, and the moon has only about $\frac{1}{73}$d part as much attracting power as the earth, it follows that her atmosphere, if she has one, ought to be much less dense than ours.

243. That no water exists upon the moon's surface,

---

241. Form of lunar mountains? Number of craters visible? Appearance of surface, as seen by Dr. Scoresby? Dr. Herschel's supposition?
242. Has the moon an atmosphere? Dr. Scoresby's statement? (Remark of author? Why moon's atmosphere must be comparatively rare?)
243. Why thought there is no water on the moon?

THE MOON.   115

has been inferred from the fact, that it would be converted into steam or vapor during her long and hot days, and also from the fact that no *clouds* are ever seen floating around her.

244. Professors Baer and Mädler, of Berlin, have constructed a map of the moon, which is characterized by Professor Nichol, of Glasgow, as " vastly more accurate than any map of the earth we can yet produce," and as "the only authentic and valuable work of the kind in existence."

The following is a list of the principal lunar mountains, with their hight, according to the recent measurements of Mädler:

	Feet.	Miles.		Feet.	Miles.
Posidonius	19,830	3·76	Clavius	19,030	3·60
Tycho	20,190	3·83	Huygens	18,670	3·54
Calippus	20,390	3·86	Blancanus	18,010	3·41
Casatus	22,810	4·32	Movetus	18,440	3·49
Newton	23,830	4·52			

245. The apparent position of the moon in the heavens is one of the principal means by which mariners ascertain their longitude at sea. So regular is her motion, that her "*place*," as viewed from any fixed point on the earth, at any specified time, and with reference to the four stars that lie in or near her, may be determined for months and years to come; and, by observing how far she appears out of place, either east or west, at the time specified, we may determine how far we are east or west of the place for which her longitude is given in the tables.

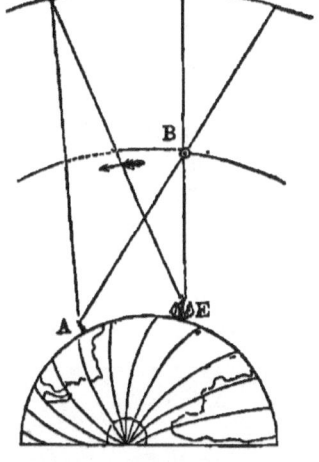

Let A in the cut represent Greenwich Observatory, near London. B is the moon, and C her apparent place among the distant stars, about 40° west of the star D. The ship E, having Greenwich time, as well as her own local time, sails from London westward; but on observing the moon when, by Greenwich time, she ought to be at C, she is found to be at F, or only about 20° west of the star D. It is therefore obvious that

---

244. What celebrated chart mentioned? How characterized? (What list of mountains given? General hight?)
245. What use made of the moon in navigation? Explain the process. What called? What other method for determining longitude?

the *ship* is *west* of Greenwich, as the moon appears *east* of her Greenwich place. From this difference between her place as laid down in the tables, and her observed place, as referred to certain prominent stars, the mariner determines how far he is east or west of the meridian of Greenwich. The moon's geocentric place (or place, as viewed from the center of the earth) may be given instead of her Greenwich place, and the same conclusions arrived at. In either case, this is called the *lunar method* of determining the longitude. It is also ascertained by simple comparison of local and standard time, as explained at 151.

246. The best time for observing the moon with a telescope is from the change to the first quarter, and from the third quarter to the change. Near the first and third quarters, the shadows of objects are seen at right angles with the line of vision, and to the best advantage; while at full moon, objects cast no shadows visible to us.

CHAPTER VI.

ECLIPSES OF THE SUN AND MOON.

247. An *Eclipse* is a partial or total obscuration or darkening of the sun or moon, by the intervention of some opake body. Eclipses are either *solar* or *lunar*. A *solar eclipse* is an eclipse of the sun, and a *lunar eclipse* is an eclipse of the moon. A solar eclipse is caused by the moon, when she passes between the earth and the sun, in her revolution eastward, and casts her shadow upon the earth. A lunar eclipse takes place when the moon is in opposition to the sun, and passes through a portion of the earth's shadow.

The general law of shadows may be illustrated by the following:

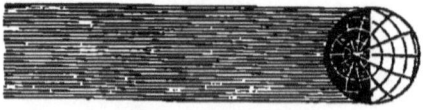 

Here the sun and planet are represented as *of the same size*, and the shadow of the latter is in the form of a *cylinder*.

---

246. When is the best time for viewing the moon with a telescope? Why?
247. What is an *eclipse*? A solar? Lunar? Cause of solar eclipses? Of lunar? When do lunar eclipses take place? (Illustrate the laws of shadows by diagram on blackboard.)

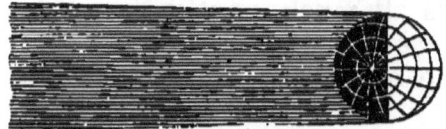 

In this cut, *the opake body is the larger*, and the shadow projected from it *diverges*, or grows more broad as the distance from the planet increases.

Here *the luminous body is the larger*, and the shadow converges to a point, and takes the form of a cone.

Here, also, the luminous body is the larger, and both precisely of the same size as in the cut preceding; but being placed *nearer* each other, the shadow is shown to be considerably shorter.

248. All the planets, both primaries and secondaries, cast shadows in a direction opposite the sun (see the adjoining cut). The form and length of these shadows depend upon the comparative magnitude of the sun and planet, and their distance from each other. If the sun and a planet were of the same size, the shadow of the planet would be in the form of a *cylinder*, whatever its distance. If the planet was *larger* than the sun, the shadow would *diverge*, as we proceed from the planet off into space; and the nearer the sun, the more divergent the shadow would be.

SHADOWS OF THE PLANETS.

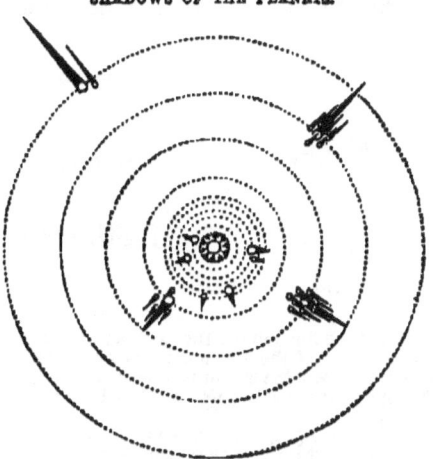

---

248. What said of the shadows of the planets? Of their *form* and *length?* How would it be if the sun and planet were of the same size? If the planet

But as the planets are all much *smaller* than the sun, the shadows all *converge* to a point, and take the form of a *cone;* and the nearer to the sun, the shorter its shadow.

<small>These principles are partly illustrated in the preceding cut. The planets nearest the sun have comparatively short shadows, while those more remote extend to a great distance. No primary, however, casts a shadow long enough to reach the next exterior planet.</small>

249. Eclipses of the sun must always happen at *New Moon*, and those of the moon at *Full Moon*. The reason of this is, that the moon can never be between us and the sun, to eclipse him, except at the time of her change, or new moon; and she can never get into the earth's shadow, to be eclipsed herself, except when she is in opposition to the sun, and it is full moon.

250. If the moon's orbit lay exactly in the plane of the ecliptic, she would eclipse the sun at every *change*, and be eclipsed herself at every *full;* but as her orbit departs from the ecliptic over 5° (216), she may pass either above or below the sun at the time of her change, or above or below the earth's shadow at the time of her full.

NEW AND FULL MOONS WITHOUT ECLIPSES.

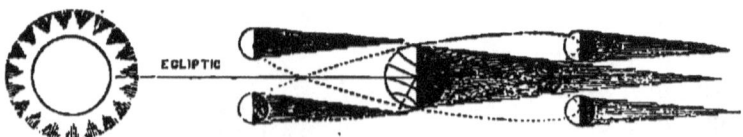

Shadow above the Earth.   Above the Earth's shadow.

Shadow below the Earth.   Below the Earth's shadow.

<small>1. Let the line joining the earth and the sun represent the plane of the ecliptic. Now as the orbit of the moon departs from this plane about 5° 9', she may appear either *above* or *below* the sun at new moon, as represented in the figure, and her shadow may fall above the north pole or below the south. At such times, then, there can be no solar eclipse.

2. On the right, the moon is shown at her full, both above and below the earth's shadow, in which case there can be no lunar eclipse.</small>

---

was largest? If brought nearer? How if planets smallest? How affected by distance? (How, then, with planets nearest the sun? More remote? Does any primary throw its shadow out to the next exterior planet?)

249. At what time of the moon do solar eclipses always occur? Lunar? Why?

250. Why not two eclipses every lunar month? (Illustrate.)

ECLIPSES OF THE SUN AND MOON.

251. Eclipses of the sun always come on from the west, and pass over eastward; while eclipses of the moon come on from the east, and pass over westward. This is a necessary result of the eastward motion of the moon in her orbit.

LUNAR ECLIPSE.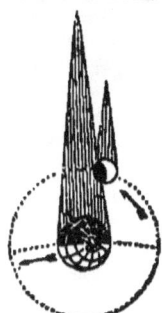

SOLAR ECLIPSE.

1. In the right hand cut, the moon is seen revolving eastward, throwing her shadow upon the earth, and hiding the western limb of the sun. In some instances, however, when the eclipse is very slight, it may first appear on the *northern* or *southern* limb of the sun—that is, the upper or lower side; but even then its *direction* must be from west to east. It will also be obvious from this figure, that the *shadow* of the moon upon the earth must also traverse her surface from west to east; consequently the eclipse will be visible earlier in the west than in the east.

2. On the left, the moon is seen striking into the earth's shadow from the west, and having her eastern limb first obscured. By holding the book up south of him, the student will see at once why the revolution of the moon eastward must cause a solar eclipse to proceed from west to east, and a lunar eclipse from east to west. To locate objects and motions correctly, the student should generally imagine himself looking to the south, as we are situated north of the equinoctial. The student should bear in mind that nearly all the cuts in the book are drawn to represent a view from northern latitude upon the earth. Hence by holding the book up *south* of him, the cuts will generally afford an accurate illustration both of the positions and motions of the bodies represented.

860,000 MILES.

252. Eclipses can never take place, except when the moon is near the ecliptic; or, in other words, at or near one of her nodes. At all other times, she passes above or below the sun, and also above or below the earth's shadow. It is not necessary that she should be *exactly* at her node, in order that an eclipse occur. If she is within 17° of her node at the time of her change, she will eclipse the sun; and if within 12° of her node at her full, she will strike into the earth's shadow, and be more or less eclipsed. These distances are called, respectively, the solar and lunar *ecliptic limits*.

---

251. What is the direction of a solar eclipse? A lunar? Why this difference?
252. Where must the moon be, with respect to the ecliptic and her nodes, in order to an eclipse? What meant by *ecliptic limits?* Name the distance of each, respectively, from the node. (Illustrate.)

This subject may be understood by consulting the following figure:

THE MOON CHANGING AT DIFFERENT DISTANCES FROM HER NODES.

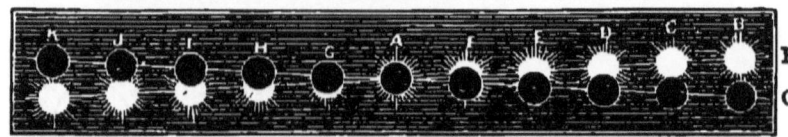

1. Let the line E E represent the ecliptic, and the line O O the plane of the moon's orbit. The light globes are the sun, and the dark ones the moon, which may be imagined as much nearer the student; hence their apparent diameter is the same.

2. Let the point A represent the node of the moon's orbit. Now if the change occur when the moon is at B, she will pass *below* the sun. If when at C, she will just touch his lower limb. At C, she will eclipse him a little, and so on to A; at which point, if the change occurs, the eclipse would be central, and probably total.

3. If the moon was at G, H, I, or J, in her orbit, when the change occurred, she would eclipse the upper or northern limb of the sun, according to her distance from her node at the time; but if she was at K, she would pass above the sun, and would not eclipse him at all. The points C and J will represent the *Solar Ecliptic Limits*.

253. All parts of a planet's shadow are not alike dense. The darkest portion is called the *umbra*, and the partial shadow the *penumbra*.

UMBRA AND PENUMBRA OF THE EARTH AND MOON.

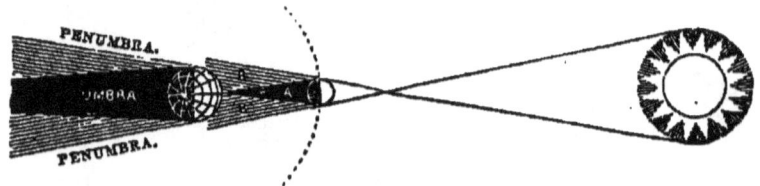

Penumbra is from the Latin *pene*, almost, and *umbra*, a shadow. In this cut, the earth's umbra and penumbra will be readily found by the lettering; while A is the umbra, and B B the penumbra, of the moon. The latter is more broad than it should be, owing to the nearness of the sun in the cut, as it never extends to much over half the earth's diameter. The student will see at once that solar eclipses can be total only to persons within the umbra; while to all on which the penumbra falls, a portion of the sun's disk will be obscured.

254. The average *length* of the earth's umbra is about 860,000 miles; and its *breadth*, at the distance of the moon, is about 6,000 miles, or three times the moon's diameter.

As both the earth and moon revolve in elliptical orbits, both the above estimates are subject to variations. The length of the earth's umbra varies from 842,217 to 871,262 miles; and its diameter where the moon passes it, varies from 5,285 to 6,365 miles.

255. The average length of the moon's umbra is about 239,000 miles. It varies from 221,148 to 252,638 miles,

---

253. What is the *umbra* of the earth or moon? The *penumbra*? (Derivation? Within which are solar eclipses total?)
254. The average length of the earth's shadow? Breadth at the moon's distance? (Do they vary? Why?)

according to the moon's distance from the sun. Its greatest diameter, at the distance of the earth, is 170 miles; but the *penumbra* may cover a space on the earth's surface 4,393 miles in diameter.

256. When the moon but just touches the limb of the sun, or the umbra of the earth, it is called an *appulse*. (See D and G, in the first cut on the opposite page.)

A *partial* eclipse is one in which only part of the sun or moon is obscured. A solar eclipse is partial to all places outside the umbra; but within the penumbra, where the whole disk is obscured, the eclipse is said to be *total*. A *central* eclipse is one taking place when the moon is exactly at one of her nodes. If *lunar*, it is *total*, as the earth's umbra is always broad enough, at the moon's distance, if centrally passed, to obscure her whole disk. But a solar eclipse may be *central* and not *total*, as the moon is not always of sufficient apparent diameter to cover the whole disk of the sun. In that case, the eclipse would be *annular* (from *annulus*, a ring), because the moon only hides the center of the sun, and leaves a bright ring unobscured.

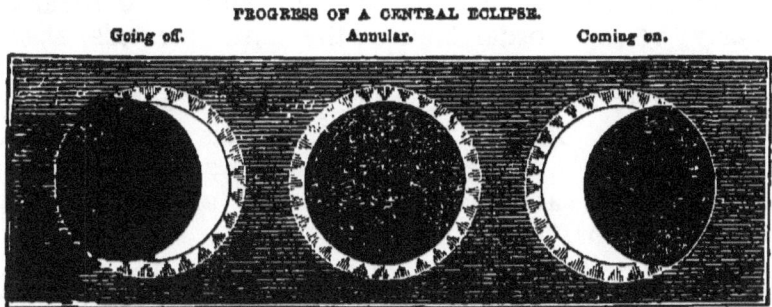

PROGRESS OF A CENTRAL ECLIPSE.
Going off.   Annular.   Coming on.

257. It has already been shown (56) that the apparent magnitudes of bodies vary as their distances vary; and as both the earth and moon revolve in elliptical orbits, it

---

255. Average length of the moon's umbra? Does it vary? Why? Greatest diameter at the earth's surface? Of penumbra?
256. What is an *appulse*? A *partial* eclipse? A *total*? A *central*? Are all central eclipses total? Why not? What called then? Why?
257. Why are some central eclipses total, and others partial and annular? (Diagram.)

follows that the moon and sun must both vary in their respective apparent magnitudes. Hence some central eclipses of the sun are total, while others are partial and annular.

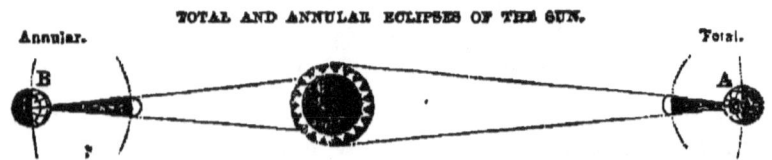

TOTAL AND ANNULAR ECLIPSES OF THE SUN.

Annular. Total.

1. At A, the earth is at her *aphelion*, and the sun being at his most distant point, will have his *least* apparent magnitude. At the same time, the moon is in *perigee*, and appears *larger* than usual. If, therefore, she pass centrally over the sun's disk, the eclipse will be *total*.

2. At B, this order is reversed. The earth is at her *perihelion*, and the moon in *apogee;* so that the sun appears *larger*, and the moon *smaller* than usual. If, then, a central eclipse occur under these circumstances, the moon will not be large enough to eclipse the whole of the sun, but will leave a ring, apparently around herself, unobscured. Such eclipse will be *annular*.

258. As the solar ecliptic's limits are further from the moon's nodes than the lunar, it results that we have more eclipses of the sun than of the moon. There may be seven in all in one year, viz., five solar and two lunar; but the most usual number is four. There can never be less than two in a year; in which case, both must be of the sun. Eclipses both of the sun and moon recur in nearly the same order, and at the same intervals, at the expiration of a cycle of 223 lunations, or 18 years of 365 days and 15 hours. This cycle is called the *Period of the Eclipses*. At the expiration of this time, the sun and the moon's nodes will sustain the same relation to each other as at the beginning, and a new cycle of eclipses begins.

259. In a total eclipse of the sun, the heavens are shrouded in darkness, the planets and stars become visible, the temperature declines, the animal tribes become agitated, and a general gloom overspreads the landscape. Such were the effects of the great eclipse of 1806. In a lunar eclipse, the moon begins to lose a portion of her

---

258. Which kind of eclipses is most frequent? Why? The greatest number in a year? How many of each? Least number, and which? Usual number? What said of the *order* of eclipses? Time of cycle?

259. Describe the effects of a total eclipse of the sun. The process of a lunar eclipse?

light and grows dim, as she enters the earth's penumbra, till at length she comes in contact with the umbra, and the real eclipse begins.

260. In order to measure and record the extent of eclipses, the apparent diameters of the sun and moon are divided into twelve equal parts, called *digits;* and in predicting eclipses, astronomers usually state which "*limb*" of the body is to be eclipsed—the southern or northern—the time of the first contact, of the nearest approach of centers, direction, and number of digits eclipsed.

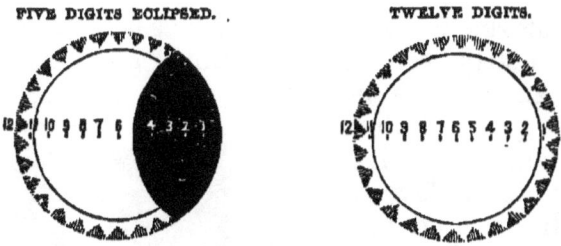

FIVE DIGITS ECLIPSED.   TWELVE DIGITS.

261. The last *annular* eclipse visible in the United States occurred May 26, 1854. The next total eclipse of the sun will be August 7, 1869.

Some of the ancients and all barbarous nations formerly regarded eclipses with amazement and fear, as supernatural events, indicating the displeasure of the gods. Columbus is said to have made a very happy use of this superstition. When the inhabitants of St. Domingo refused to allow him to anchor, in 1502, or to furnish him supplies, he told them the Great Spirit was offended at their conduct, and was about to punish them. In proof, he said the moon would be darkened *that very night;* for he knew an eclipse was to occur. The artifice led to a speedy and ample supply of his wants.

262. Eclipses can be calculated with the greatest precision, not only for a few years to come, but for centuries

---

260. How are eclipses measured and recorded?
261. When the next annular eclipse visible in this country? The next total? How have the ignorant and superstitious regarded eclipses? Anecdote of Columbus?

and ages either past or to come. This fact demonstrates the truth of the Copernican theory, and illustrates the *order* and *stability* that everywhere reign throughout the planetary regions.

CHAPTER VII.

SATELLITES OF THE EXTERIOR PLANETS.

263. JUPITER is attended by four satellites or moons. They are easily seen with a common spy-glass, appearing like small stars near the primary. (See adjoining cut, and note at 178.) By watching them for a few evenings, they will be seen to change their places, and to occupy different positions. At times, only one or two may be seen, as the others are either between the observer and the planet, or *beyond* the primary, or eclipsed by his shadow.

TELESCOPIC VIEWS OF THE MOONS OF JUPITER.

264. The size of these satellites is about the same as our moon, except the second, which is a trifle less. The first is about the distance of our moon; and the others, respectively, about two, three, and five times as far off.

COMPARATIVE DISTANCES OF JUPITER'S MOONS.

4th.   3d.   2d.   1st.

---

262. What said of the calculation of eclipses? What does this demonstrate and illustrate?
263. How many moons has Jupiter? How seen? Why not all seen at once?
264. Their size? Distances? Periods? Why so rapid?

Their periods of revolution are from 1 day 18 hours to 17 days, according to their distances. This rapid motion is necessary, in order to counterbalance the powerful centripetal force of the planet, and to keep the satellites from falling to his surface.

The magnitudes, distances, and periods of the moons of Jupiter are as follows:

	Diameter in miles.	Distance.	Periodic times.
1st	2,500	259,000	1 day 18 hours.
2d	2,068	414,000	3 " 12 "
3d	3,377	647,000	7 " 14 "
4th	2,890	1,164,000	17 " 0 "

265. The orbits of Jupiter's moons are all in or near the plane of his equator; and as his orbit nearly coincides with the ecliptic, and his equator with his orbit, it follows that, like our own moon, his satellites revolve near the plane of the ecliptic. On this account, they are sometimes between us and the planet, and sometimes beyond him, and seem to oscillate, like a pendulum, from their greatest elongation on one side to their greatest elongation on the other.

266. Their direction is from west to east, or in the direction their primary revolves, both upon his axis and in his orbit. From the fact that their elongations east and west of Jupiter are nearly the same at every revolution, it is concluded that their orbits are but slightly elliptical. They are supposed to revolve on their respective axes, like our own satellite, the moon, once during every periodic revolution.

267. As these orbits lie near the plane of the ecliptic, they have to pass through his broad shadow when in opposition to the sun, and be totally eclipsed at every revolution. To this there is but one exception. As the fourth satellite departs about 3° from the plane of Jupiter's orbit, and is quite distant, it sometimes passes *above* or *below* the shadow, and escapes eclipse. But such escapes are not frequent.

---

265. How are their orbits situated? How satellites appear to move?
266. Direction of secondaries? Form of orbits? How ascertained? What motion on axes?
267. What said of eclipses? Of fourth satellite? Of solar eclipses upon Jupiter? Number of solar and lunar?

These moons are not only often eclipsed, but they often eclipse Jupiter, by throwing their own dark shadows upon his disk. They may be seen like dark round spots traversing it from side to side, causing, whenever that shadow falls, an eclipse of the sun. Altogether, about forty of these eclipses occur in the system of Jupiter every month.

268. The *immersions* and *emersions* of Jupiter's moons have reference to the phenomena of their being eclipsed. Their *entrance into* the shadow is the *immersion;* and their *coming out of it* the *emersion.*

ECLIPSES OF JUPITER'S MOONS, EMERSIONS, ETC.

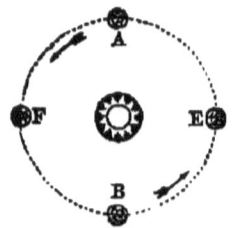 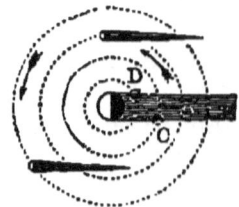

1. The above is a perpendicular view of the orbits of Jupiter's satellites. His broad shadow is projected in a direction opposite the sun. At O, the second satellite is suffering an *immersion*, and will soon be totally eclipsed; while at D, the first is in the act of *emersion*, and will soon appear with its wonted brightness. The other satellites are seen to cast their shadows off into space, and are ready in turn to eclipse the sun, or cut off a portion of his beams from the face of the primary.

2. If the earth were at A in the cut, the *immersion*, represented at O, would be invisible; and if at B, the *emersion* at D could not be seen. So, also, if the earth were exactly at F, neither could be seen; as Jupiter and all his attendants would be directly beyond the sun, and would be hid from our view.

269. The system of Jupiter may be regarded as a miniature representation of the solar system, and as furnishing triumphant evidence of the truth of the Copernican theory. It may also be regarded as a great *natural clock*, keeping absolute time for the whole world; as the immersions and emersions of his satellites furnish a uniform standard, and, like a vast chronometer hung up in the heavens, enable the mariner to determine his longitude upon the trackless deep.

---

268. What are the *immersions* and *emersions* of Jupiter's moons? (Are the immersions and emersions always visible from the earth? Why not? Illustrate.)

269. How may the system of Jupiter be regarded? What use made of in navigation? (Illustrate method. Much used?)

SATELLITES OF THE EXTERIOR PLANETS. 127

By long and careful observations upon these satellites, astronomers have been able to construct tables, showing the exact time when each immersion and emersion will take place, at Greenwich Observatory, near London. Now suppose the tables fixed the time for a certain satellite to be eclipsed at 12 o'clock at Greenwich, but we find it to occur at 9 o'clock, for instance, by our local time; this would show that our time was three hours *behind* the time at Greenwich; or, in other words, that we were three hours, or 45°, *west* of Greenwich. If our time was *ahead* of Greenwich time, it would show that we were *east* of that meridian, to the amount of 15° for every hour of variation. But this method of finding the longitude is less used than the "lunar method" (Art. 245), on account of the greater difficulty of making the necessary observations.

270. By observations upon the eclipses of Jupiter's moons, as compared with the tables fixing the time of their occurrence, it was discovered that light had a progressive motion, at the rate of about 200,000 miles per second.

1. This discovery may be illustrated by again referring to the opposite cut. In the year 1675, it was observed by Roemer, a Danish astronomer, that when the earth was nearest to Jupiter, as at E, the eclipses of his satellites took place 8 minutes 18 seconds *sooner* than the mean time of the tables; but when the earth was farthest from Jupiter, as at F, the eclipses took place 8 minutes and 13 seconds *later* than the tables predicted the entire difference being 16 minutes and 26 seconds. This difference of time he ascribed to the progressive motion of light which he concluded required 16 minutes and 26 seconds to cross the earth's orbit from E to F.

2. This progress may be demonstrated as follows:—16m. 26s. = 986s. If the radius of the earth's orbit be 95 millions of miles, the diameter must be twice that, or 190 millions. Divide 190,000,000 miles by 986 seconds, and we have $192,697\frac{37}{136}$ miles as the progress of light in each second. At this rate, light would pass nearly eight times around the globe at every tick of the clock, or nearly 500 times every minute!

SATURN.

271. The moons of Saturn are eight in number, and are seen only with telescopes of considerable power. The best time for observing them is when the planet is at his equinoxes, and his rings are nearly invisible.

SATELLITES OF SATURN.

In January, 1849, the author saw five of these satellites, as represented in the adjoining cut. The rings appeared only as a line of light, extending each way from the planet, and the satellites were in the direction of the line, at different distances, as here represented.

272. These satellites all revolve eastward with the rings of the planet, in orbits nearly circular, and, with the exception of the eighth, in the plane of the rings. Their mean distances, respectively, from the planet's cen-

---

270. What discovery by observing these eclipses? (Illustrate method. Diagram. Demonstration.)
271. Number of Saturn's moons? How seen? Best time?
272. How revolve? Shape of orbits? How situated? Distances? Periods?

ter are from 123,000 to 2,366,000 miles; and their periods from 22 hours to 79 days, according to their distances.

The distances and periods of the satellites of Saturn are as follows:

Distance in miles	Periodic times	Distance in miles	Periodic times
1st......123,000	0 day 22 hours.	5th......351,000	4 days 12 hours.
2d......158,000	1 " 8 "	6th......811,000	15 " 22 "
3d......196,000	1 " 21 "	7th....2,366,000	79 " 7 "
4th......251,000	2 " 17 "		

COMPARATIVE DISTANCES OF THE MOONS OF SATURN.

273. The most distant of these satellites is the largest, supposed to be about the size of Mars; and the remainder grow smaller as they are nearer the primary. They are seldom eclipsed, on account of the great inclination of their orbits to the ecliptic, except twice in thirty years, when the rings are edgewise toward the sun. The eighth satellite, which has been studied more than all the rest, is known to revolve once upon its axis during every periodic revolution; from which it is inferred that they all revolve on their respective axes in the same manner.

1. Let the line A B represent the plane of the planet's orbit, C D his axis, and E F the plane of his rings. The satellites being in the plane of the rings, will revolve *around* the shadow of the primary, instead of passing through it, and being eclipsed.
2. At the time of his equinoxes, however, when the rings are turned toward the sun (see A and E, cut, page 92), they must be in the center of the shadow on the opposite side; and the moons, revolving in the plane of the rings, must pass through the shadow at every revolution. The eighth, however, may sometimes escape, on account of his departure from the plane of the rings, as shown in the cut.

SYSTEM OF SATURN—NO ECLIPSES.

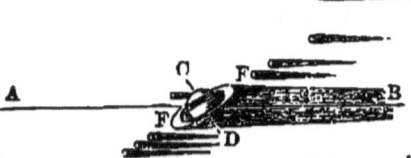

URANUS.

274. Uranus is supposed to be attended by six secondaries. Sir Wm. Herschel recorded that he saw this number, and computed their periods and distances; and on his authority the opinion is generally received, though

---

273. Size? Eclipses of? When? Why not oftener? (Illustrate.)
274. Satellites of Uranus? Upon what authority? Distances? Periods? Situation of orbits? Form? Direction in revolution? Remark of Dr. Herschel?

no other observer has ever been able to discover more than three. They are situated at various distances, and revolve in from 1 day and 21 hours to 117 days. Their orbits are nearly perpendicular to the ecliptic, and they revolve *backward*, or from east to west, contrary to all the other motions of our planetary system. Their orbits are nearly circular, and they are described by Dr. Herschel as "the most difficult objects to obtain a sight of, of any in our system."

The distances and periods of the system of Uranus, as laid down by Dr. Herschel, are as follows:

	Distance in miles.	Periodic times.		Distance in miles.	Periodic times.
1st	224,000	5 days 21 hours.	4th	390,000	13 days 11 hours.
2d	296,000	8 " 17 "	5th	777,000	38 " 2 "
3d	340,000	10 " 23 "	6th	1,556,000	117 " 17 "

NEPTUNE.

275. Neptune is known to be attended by one satellite, and suspected of having two. Professor Bond, of Cambridge, Mass., states that he has at times been quite confident of seeing a *second*. The mean distance of the known satellite from its primary is 230,000 miles, or near the distance of our moon. Its period is only 5 days and 21 hours.

We have here another illustration of the great law of planetary motion explained at 74. So great is the attractive power of Neptune, that to keep a satellite, at the distance of our moon, from falling to his surface, it must revolve some five times as swiftly as she revolves around the earth. The centripetal and centrifugal forces must be balanced in all cases, as the laws of gravitation and planetary motion, discovered by Newton and Kepler, extend to and prevail among all the secondaries.

CHAPTER VIII.

NATURE AND CAUSE OF TIDES.

276. TIDES are the alternate rising and falling of the waters of the ocean, at regular intervals. *Flood tide* is when the waters are *rising;* and *ebb* tide, when they are

275. What said of Neptune's secondaries? Remark of Prof. Bond? Distance and period of the known satellite? (Remark in note.)
276. What are tides? Flood and ebb tides? High and low? How often do they ebb and flow?

falling. The highest and lowest points to which they go are called, respectively, *high* and *low* tides. The tides ebb and flow twice every twenty-four hours—*i. e.*, we have two flood and two ebb tides in that time.

277. The tides are not uniform, either as to time or amount. They occur about 50 minutes later every day (as we shall explain hereafter), and sometimes rise much higher and sink much lower than the average. These extraordinary high and low tides are called, respectively, *spring* and *neap* tides.

278. The *cause* of the tides is the attraction of the sun and moon upon the waters of the ocean. But for this foreign influence, as we may call it, the waters having found their proper level, would cease to heave and swell, as they now do, from ocean to ocean, and would remain calm and undisturbed, save by its own inhabitants and the winds of heaven, from age to age.

NO TIDE.

In this figure, the earth is represented as surrounded by water, in a state of rest or equilibrium, as it would be were it not acted upon by the sun and moon.

279. To most minds, it would seem that the natural effect of the moon's attraction would be to produce a single tide-wave on the side of the earth toward the moon. It is easy, therefore, for students to conceive how the moon can produce one flood and one ebb tide in twenty-four hours.

ONE TIDE-WAVE.

1. In this cut, the moon is shown at a distance above the earth, and attracting the waters of the ocean, so as to produce a high tide at A. But as the moon makes her apparent westward revolution around the earth but once a day, the simple raising of a flood tide on the side of the earth toward the moon, would give us but one flood and one ebb tide in twenty-four hours; whereas it is known that we have two of each.

2. "The tides," says Dr. Herschel, "are a subject on which many persons find a strange difficulty of conception. That the moon, by her attraction, should heap up the waters of the ocean under her, seems to many persons very natural. That the same cause should, at the same time, heap them up on the opposite side of the earth (viz., at B in the figure), seems to many palpably absurd. Yet nothing is more true."

280. Instead of a single tide-wave upon the waters of

---

277. Are the tides uniform? What variation of time? As to amount? What are these extraordinary high and low tides called?
278. The *cause* of tides? How but for this influence?
279. What most obvious effect of the moon's attraction? (Substance of note 1? Remark of Dr. Herschel?)

NATURE AND CAUSE OF TIDES. 131

the globe, directly under the moon, it is found that on the side of the earth directly opposite there is another high tide; and that half way between these two high tides are two low tides. These four tides, viz., two high and two low, traverse the ocean from east to west every day, which accounts for both a flood and an ebb tide every twelve hours.

TWO TIDE-WAVES.

In this cut, we have a representation of the tide-waves as they actually exist, except that their hight, as compared with the magnitude of the earth, is vastly too great. It is designedly exaggerated, the better to illustrate the principle under consideration. While the moon at A attracts the waters of the ocean, and produces a high tide at B, we see another high tide at C on the opposite side of the globe. At the same time it is low tide at D and E.

281. The principal cause of the tide-wave on the side of the earth opposite the moon is the *difference* of the moon's attraction on different sides of the earth.

If the student well understands the subject of gravitation (65), he will easily perceive how a difference of attraction, as above described, would tend to produce an elongation of the huge drop of water called the earth. The diameter of the earth amounts to about $\frac{1}{30}$th of the moon's distance; so that, by the rule (69), the difference in her attraction on the side of the earth toward her, and the opposite side, would be about $\frac{1}{T}$th. The attraction being stronger at B (in the last cut) than at the earth's center, and stronger at her center than at C, would tend to *separate* these three portions of the globe, giving the waters an elongated form, and producing two opposite tide-waves, as shown in the cut.

282. A secondary cause of the tide-wave on the side of the earth opposite the moon, is the revolution of the earth around the common center of gravity between the earth and moon, thereby generating an increased centrifugal force on that side of the earth.

The center of gravity between the earth and moon is the point where they would exactly balance each other, if connected by a rod, and poised upon a fulcrum.

CENTER OF GRAVITY BETWEEN THE EARTH AND MOON.

This point, which, according to Ferguson, is about 6,000 miles from the earth's center, is represented at A in the above, and also in the next cut.

---

280. How many tide-waves are there on the globe, and how situated?
281. State the principal cause of the wave opposite the moon? (Demonstrate by diagram.)
282. What other cause operates with the one just stated to produce the tide-wave opposite the moon? (What is the center of gravity between the earth and the moon? Where is it situated? Illustrate the operation of this secondary cause. Diagram.)

SECONDARY CAUSE OF HIGH TIDE OPPOSITE THE MOON.

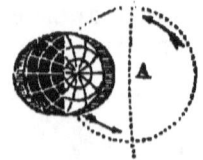 

1. The point A represents the center of gravity between the earth and moon; and as is this point which traces the regular curve of the earth's orbit, it is represented in the arc of that orbit, while the earth's center is 6,000 miles one side of it. Now the law of gravitation requires that while both the moon and earth revolve around the sun, they should also revolve around the common center of gravity between them, or around the point A. This would give the earth a *third revolution*, in addition to that around the sun and on her axis. The small circles show her path around the center of gravity, and the arrows her direction.

2. This motion of the earth would slightly increase the centrifugal tendency at B, and thus help to raise the tide-wave opposite the moon. But as this motion is slow, corresponding with the revolution of the moon around the earth, the centrifugal force could not be greatly augmented by such a cause.

283. As the moon, which is the principal cause of the tides, is revolving eastward, and comes to the meridian later and later every night, so the tides are about 50 minutes later each successive day. This makes the interval between two successive high tides 12 hours and 25 minutes. Besides this daily lagging with the moon, the highest point of the tide-wave is found to be about 45° behind or east of the moon, so that high tide does not occur till about three hours after the moon has crossed the meridian. The waters do not at once yield to the impulse of the moon's attraction, but continue to rise after she has passed over.

TIDE-WAVES BEHIND THE MOON.
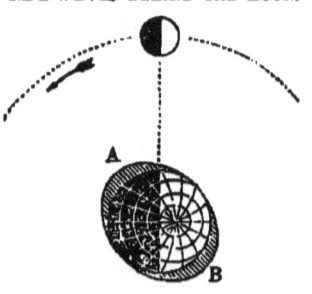

In the cut, the moon is on the meridian, but the highest point of the wave is at A, or 45° east of the meridian; and the corresponding wave on the opposite side at B is equally behind.

284. The time and character of the tides are also affected by winds, and by the situation of different places. Strong winds may either retard or hasten the tides, or may increase or diminish their hight; and if a place is situated on a large bay, with but a narrow opening into the sea, the tide will be longer in rising, as the bay has

283. What daily lagging of the tides? Interval between two successive high tides? What other lagging? Cause of this last?
284. What modification of the time and character of the tides?

to fill up through a narrow gate. Hence it is not usually high tide at New York till eight or nine hours after the moon has passed the meridian.

285. As both the sun and moon are concerned in the production of tides, and yet are constantly changing their positions with respect to the earth and to each other, it follows that they sometimes *act against each other*, and measurably neutralize each other's influence; while at other times they *combine* their forces, and mutually assist each other. In the latter case, an unusually high tide occurs, called the *Spring Tide*. This happens both at new and full moon.

CAUSE OF SPRING TIDES.

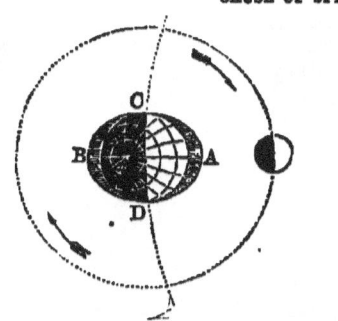

1. Here the sun and moon, being in conjunction, unite their forces to produce an extraordinary tide. The same effect follows when they are in opposition; so that we have two spring tides every month—namely, at new and full moon.
2. If the tide-waves at A and B are one-third *higher* at the moon's quadrature than usual, those of C and D will be one-third *lower* than usual.

286. Although the sun attracts the earth much more powerfully, as a whole, than the moon does, still the moon contributes more than the sun to the production of tides. Their relative influence is as *one* to *three*. The nearness of the moon makes the *difference* of her attraction on different sides of the earth much greater than the difference of the sun's attraction on different sides.

It must not be forgotten that the tides are the result not so much of the attraction of the sun and moon, *as a whole*, as of the *difference* in their attraction on different sides

---

285. Do the sun and moon always act together in attracting the waters? Why not? How affect each other's influence? Effect on the tides? What are *Spring Tides*? When do they occur? (Illustrate by diagram the cause of spring tide, when the sun and moon are in conjunction.)

286. Comparative influence of sun and moon in the production of tides? Why moon's influence the greatest? (Substance of note? Demonstration.)

of the earth, caused by a difference in the *distances* of the several parts. The attraction being inversely, as the square of the distance (69), the influence of the sun and moon, respectively, must be in the ratio of the earth's diameter to their distances. Now the difference in the distance of two sides of the earth from the moon is $\frac{1}{30}$ th of the moon's distance; as $240{,}000 \div 8{,}000 = 30$; while the difference, as compared with the distance of the sun, is only $\frac{1}{11875}$ th, as $95{,}000{,}000 \div 8{,}000 = 11{,}875$.

287. When the moon is in *quadrature*, and her influence is partly neutralized by the sun, which now acts against her, the result is a very low tide, called *Neap Tide*.

SPRING AND NEAP TIDES.

The whole philosophy of spring and neap tides may be illustrated by the annexed diagram.

1. On the right side of the cut, the sun and moon are in *conjunction*, and unite to produce a *spring tide*.
2. At the first quarter, their attraction acts at right angles, and the sun, instead of contributing to the lunar tide-waves, detracts from it to the amount of his own attractive force. The tendency to form a tide of his own, as represented in the figure, *reduces* the moon's wave to the amount of one-third.
3. At the full moon, she is in *opposition* to the sun, and their joint attraction acting again in the same line, tends to elongate the fluid portion of the earth, and a second *spring tide* is produced.
4. Finally, at the third quarter, the sun and moon act *against* each other again, and the second neap tide is the result. Thus we have two spring and two neap tides during every lunation—the former at the moon's syzygies, and the latter at her quadratures.

288. The tides are subject to another periodic variation, caused by the declination of the sun and moon north and south of the equator. As the tendency of the tide-wave is to rise directly under the sun and moon, when they are in the south, as in winter, or in the north, as in summer, every alternate tide is higher than the intermediate one.

TIDES AFFECTED BY DECLINATION.

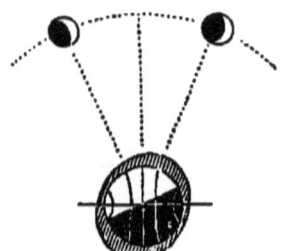

At the time of the equinoxes, the sun being over the equator, and the moon within 5¼° of it, the crest of the great tide-wave will be on the equator; but as the sun and moon decline south to A, one tide-wave forms in the south, as at B, and the opposite one in the north, as at C. If the declination was *north*, as shown at D, the order of the tides would be reversed. The following diagram,

287. What are *Neap Tides?* Their cause? (Illustrate entire philosophy by diagram.)
288. What other periodic variations mentioned? (Explain cause, and illustrate.)

NATURE AND CAUSE OF TIDES. 135

If carefully studied, will more fully illustrate the subject of the alternate high and low tides, in high latitudes, in winter and summer:

ALTERNATE HIGH AND LOW TIDES.

1. Let the line A A represent the plane of the *ecliptic*, and B B the *equinoctial*. On the 21st of June, the day tide-wave is north, and the evening wave south, so that the tide following about three hours after the sun and moon will be higher than the intermediate one at 8 o'clock in the morning.
2. On the 23d of December, the sun and moon being over the southern tropic, the highest wave in the southern hemisphere will be about 8 o'clock P. M., and the lowest about 8 o'clock A. M.; while at the north, this order will be reversed. It is on this account that in high latitudes every alternate tide is higher than the intermediate ones; the evening tides in summer exceeding the morning tides, and the morning tides in winter exceeding those of evening.

289. All spring and neap tides are not alike as to their elevation and depression. As the *distances* of the sun and moon are varied, so are the tides varied, especially by the variations of the moon.

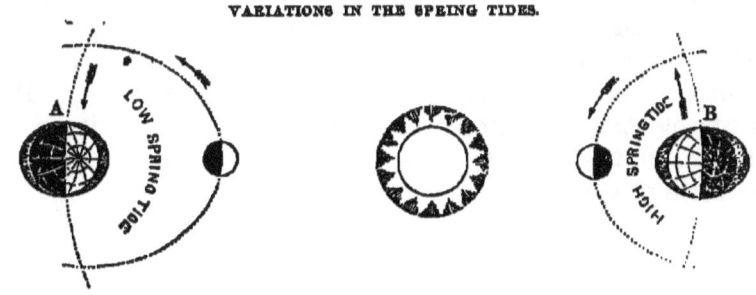

VARIATIONS IN THE SPRING TIDES.

1. At A, the earth is in *aphelion*, and the moon in *apogee*. As both the sun and moon are at their greatest distances, the earth is least affected by their attraction, and the spring tides are proportionately low.
2. At B, the earth is in *perihelion*, and the moon in *perigee;* so that both the sun and moon exert their greatest influence upon our globe, and the spring tides are highest, as shown in the figure. In both cases, the sun and moon are in conjunction, but the variation in the *distances* of the sun and moon causes variations in the spring tides.

290. In the open ocean, especially the Pacific, the tide rises and falls but a few feet; but when pressed into narrow bays or channels, it rises much higher than under ordinary circumstances.

---

289 Are all spring and neap tides alike? By what are they modified? (Illustrate by diagram.)
290. Hight of tides in open seas? How in narrow bays and channels? (Hight at different points on our coast?)

136   ASTRONOMY.

The average elevation of the tide at several points on our coast is as follows:

Cumberland, head of the Bay of Fundy.............................. 71 feet.
Boston.. 11¼ "
New Haven.. 8 "
New York.. 5 "
Charleston, S. C.. 6 "

291. As the great tide-waves proceed from east to west, they are arrested by the continents, so that the waters are permanently higher on their east than on their west sides. The Gulf of Mexico is 20 feet higher than the Pacific Ocean, on the other side of the Isthmus; and the Red Sea is 30 feet higher than the Mediterranean. Inland seas and lakes have no perceptible tides, because they are too small, compared with the whole surface of the globe, to be sensibly affected by the attraction of the sun and moon.

We have thus stated the principal facts connected with this complicated phenomenon, and the *causes* to which they are generally attributed. And yet it is not certain that the philosophy of tides is to this day fully understood. *La Place*, the great French mathematician and astronomer, pronounced it one of the most difficult problems in the whole range of celestial mechanics. It is probable that the atmosphere of our globe has its tides, as well as the waters; but we have no means, as yet, for definitely ascertaining the fact.

CHAPTER IX.

OF COMETS.

292. COMETS are a singular class of bodies, belonging to the solar system, distinguished for their long trains of light, their various shapes, and the great eccentricity of their orbits. Their name is from the Greek *coma*, which

---

291. Direction of tide-waves? What result? Instances cited? Have inland seas and lakes any tides? Why not? Remarks respecting philosophy of tides? Of La Place? Atmospheric tides?

292. What are comets? Derivation of name? Are they opake or self-luminous?

signifies *beard* or *hair*, on account of their bearded or hairy appearance. They are known to be opake, from the fact that they sometimes exhibit phases, which show that they shine only by reflection.

293. Comets usually consist of three parts—the nucleus, the envelope, and the tail. The *nucleus* is what may be called the *body* or *head* of the comet. The *envelope* is the nebulous or hairy covering that surrounds the nucleus; and the *tail* is the expansion or elongation of the envelope. But all comets have not these parts. Some have no perceptible nucleus; their entire structure being like that of a thin vapory cloud passing through the distant heavens. Others have but a slight envelope around a strongly marked nucleus.

GREAT COMET OF 371 BEFORE CHRIST.

The great comet that appeared 371 years before Christ exhibited the different parts of a comet with great distinctness; on which account, as well as for its striking magnificence, we give a view of it in the above cut.

294. The *tails* of comets usually lie in a direction *opposite to the sun*, so that from perihelion to aphelion they *precede* their nuclei or heads; or, in other words, comets seem, after having passed their perihelion, to *back out* of the solar system. Their tails are usually *curved* more or less, being concave toward the region from whence they come. This is well shown in the comets of 1811, 1843, and in the following cut. That of 1689 is said to have been curved like a Turkish sabre. The *cause* of this curvature of the tails of comets is supposed to be a very rare ethereal substance which pervades

---

293. Parts of a comet? Describe each. Have all comets these three parts? (What comet shown as a sample in the cut?)
294. Direction of the tails of comets? How curved? Cause of this curvature?

space, and offers a slight resistance to their progress. Of course it must be almost infinitely attenuated, as the comets themselves are a mere vapor, which could make no progress through the spaces of the heavens, were they not very nearly a vacuum. They could no more pass a medium as dense as our atmosphere, than an ordinary cloud could pass through the waters of the sea.

295. The form of the comets' *orbits* is generally that of an ellipse greatly flattened or elongated. The sun being near one end of the ellipse, and the planets comparatively in his immediate neighborhood, the comets are in the vicinity of the sun and planets but a short time, and then hasten outward again beyond the limits of human vision, with the aid of the best telescopes, to be gone again for centuries.

ORBIT OF A COMET.

Here it will be seen that the orbit is very eccentric, that the perihelion point is very near the sun, and the aphelion point very remote.

296. The *tails* of comets *do not continue of the same uniform length*. They *increase* both in length and breadth as they approach the sun, and *contract* as they recede from him, until they often nearly disappear before the comet gets out of sight. Instances have occurred in which tails of comets have been *suddenly expanded* or *elongated* to a great distance. This is said to have been the case with the great comet of 1811.

---

295. Form of the *orbits* of comets? What near the earth but little of the time?

296. What said of the contraction and expansion of the tails of comets? What specimen shown in the cut?

GREAT COMET OF 1811.

297. Comets have been known to exhibit *several tails* at the same time. That of 1744, represented in the cut, had no less than *six tails* spread out in the heavens, like an enormous fan. The comet of 1823 is said to have had *two tails*, one of which extended *toward* the sun.

COMET OF 1744.

The comet of 1744, represented in this cut, excited great attention and interest. It exhibited no train till within the distance of the orbit of Mars from the sun; but early in March it appeared with a tail divided into six branches, all diverging, but curved in the same direction. Each of these tails was about 4° wide, and from 30° to 44° in length. The edges were bright and decided, the middle faint, and the intervening spaces as dark as the rest of the firmament, the stars shining in them. When circumstances were favorable to the display of this remarkable body, the scene was striking and magnificent, almost beyond description.

298. The *heads* or *nuclei* of comets are *comparatively small*. The following table shows the estimated diameter in five different instances:

The comet of 1778 . diameter of head 33 miles.
    "   1805 .   "   "   36   "
    "   1799 .   "   "   462   "
    "   1807 .   "   "   666   "
    "   1811 .   "   "   428   "

297. Have they ever more than one tail? What peculiarity of the comet of 1823? (What specimen of comet with several tails? Describe.)

Many comets have simply the envelope, *without any tail* or elongation. Such were those that appeared in 1585 and 1763, the former of which is represented in the adjoining cut. Cassini describes the comet of 1682 as being *as round and as bright as Jupiter, without even an envelope.* But these are very rare exceptions to the general character of cometary bodies.

COMET OF 1585.

299. The *tails* of comets are often of enormous length and magnitude. That of 371 before Christ was 60° long, covering one-third of the visible heavens. In 1618, a comet appeared, which was 104° in length. Its tail had not all risen when its head reached the middle of the heavens. That of 1680 had a tail 70° long; so that though its head set soon after sundown, its tail continued visible all night.

GREAT COMET OF 1843.

The following table will show the length of the tails of some of the most remarkable comets, both in degrees and in miles. They will be characterized only by the year when they appeared:

		Deg.	Miles.		Deg.	Miles.
B. C.	371	60	140,000,000	A.D. 1744	30	35,000,000
A. D.	1456	60	70,000,000	" 1769	90	48,000,000
"	1618	104	65,000,000	" 1811	23	132,000,000
"	1680	70	123,000,000	" 1843	60	130,000,000
"	1689	68	100,000,000			

298. What of the *size* of the *nuclei* of comets? Give a few examples. What comets without tails? What specimen in the cut? What said of the comet of 1682? Are such comets numerous?

299. What of the size of the *tails* of comets? That of 371 B. C.? Of A. D. 1618? Of 1680? (What specimen in cut, and its length? State the length of some others in miles.)

300. The *velocity* with which comets often move is truly wonderful. Their motions are accelerated as they approach, and retarded as they recede from the sun; so that their velocity is greatest while passing their perihelions. The comet of 1472 described an arc of the heavens of 120° in extent in a single day! That of 1680 moved, when near its perihelion, at the rate of 1,000,000 miles per hour.

301. The *temperature* of some comets, when nearest the sun, must be very great. That of 1680 came within 130,000 miles of the sun's surface, and must have received 28,000 times the light and heat which the earth receives from the sun—a heat more than 2,000 times as great as that of red-hot iron! What substance can a comet be composed of to endure the extremes of heat and cold to which it is subject? Some have supposed that their tails were caused by the sun's light and heat rarefying and driving back the vapory substance composing the envelope.

302. The *periods* of but few comets are known. That of 1818, called *Encke's Comet*, has a period of only 3⅓ years. *Biela's Comet* has a period of 6¾ years. That of 1682 (then first noticed with care, and identified as the same that had appeared in 1456, 1531, and 1607) has a period of about 76 years. It is called *Halley's Comet*, after Dr. Halley, who determined its periodic time. The

ENCKE'S COMET.

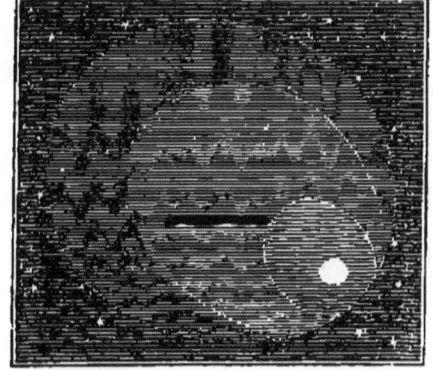

great comet of 1680 has a periodic time of 570 years, so that its next return to our system will be in the year

---

300. Velocity of comets? Uniform or not? Comet of 1472? Of 1680?
301. *Temperature?* Comet of 1680? Supposed cause of their tails?
302. Periods? Encke's? Biela's? Halley's? That of 1680? Supposed periods of others? Opinions of Prof. Nichol and Dr. Herschel?

2250. Many are supposed to have periods of thousands of years; and some have their orbits so modified by the attraction of the planets, as to pass off in parabolic curves, to return to our system no more.

Prof. Nichol is of opinion that the greater number visit our system but once, and then fly off in nearly straight lines till they pass the center of attraction between the solar system and the fixed stars, and go to revolve around other suns in the far distant heavens. Sir John Herschel expresses the same sentiment.

303. The *distances* to which those comets that return must go, to be so long absent, must be very great. Still their bounds are set by the great law of gravitation, for were they to pass the point "where gravitation turns the other way," they would never return. But some, at least, *do* return, after their "long travel of a thousand years." What a sublime conception this affords us of the almost infinite space between the solar system and the fixed stars.

ORBIT OF HALLEY'S COMET.

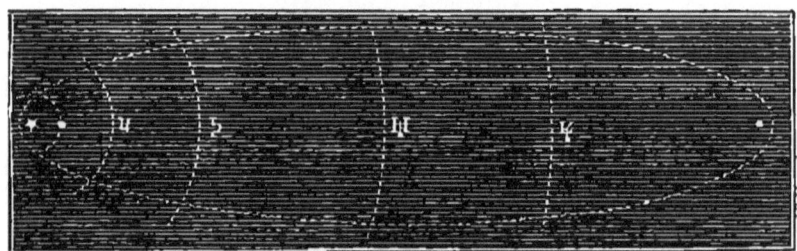

304. The *perihelion distances* of the various comets that have appeared, and whose elements have been estimated by astronomers, are also exceedingly variable. While some pass very near the sun, others are at an immense distance from him, even at their perihelion. Of 137 that have been particularly noticed,

30 passed between the sun and the orbit of Mercury.
44 between the orbits of Mercury and Venus.
34   "   "   Venus and the earth.
23   "   "   the earth and Mars.
6   "   "   Mars and Jupiter.

---

303. *Distances* to which they go? Remark respecting the law of gravitation? What specimen of orbit given?
304. What said of *perihelion* distances? How many noticed? Where did

The orbit of Encke's comet is wholly within the orbit of Jupiter, while that of Biela's extends but a short distance beyond it. The aphelion distance of Halley's comet is 3,400 millions of miles, or 550 millions of miles beyond the orbit of Neptune. But these are all comets of short periods.

ORBITS OF SEVERAL COMETS.

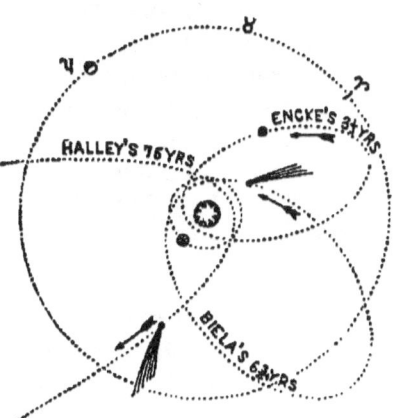

305. The *number* of comets belonging to, or that visit the solar system, is very great. Some have estimated them at several millions. When we consider that most comets are seen only through telescopes—an instrument of comparatively modern date—and that, notwithstanding this, some 450 are mentioned in ancient annals and chronicles, as having been seen with the naked eye, it is probable that the above opinion is by no means extravagant. It is supposed that not less than 700 have been seen at different times since the birth of Christ. The paths of only about 140 have been determined.

The extreme difficulty of observing comets whose nearest point is beyond the orbit of Mars, is supposed to account for the comparatively small number that have been seen without that limit; and the proximate uniformity of the distribution of their orbits over the space included within the orbit of Mars, seems to justify the conclusion, that though seldom detected beyond his path, they are nevertheless equally distributed through all the spaces of the solar heavens. Reasoning upon this hypothesis, Professor Arago concludes that there are probably *seven millions* of comets that belong to or visit the solar system.

306. The *directions* of comets are as variable as their forms or magnitudes. They enter the solar system from all points of the heavens. Some seem to come up from the immeasurable depths below the ecliptic, and, having doubled "heaven's mighty cape," again plunge downward with their fiery trains, and are lost for ages in the ethereal void. Others appear to come down from the zenith of the universe, and, having passed their peri-

they pass? (What samples given in cut? Where does the orbit of Encke's comet lie? Of Biela's? Of Halley's?)

305. The *number* of comets? What estimate? Why probably correct? How many supposed to have been seen since the birth of Christ? (Why so few seen? How supposed to be distributed? What conclusion of Arago?)

306. Direction of comets? (Remark of late writer?)

helion, reascend far above all human vision. Others again are dashing through the solar system, in all possible directions, apparently without any prescribed path, or any guide to direct them in their eccentric wanderings. Instead of revolving uniformly from east to west, like the planets, their motions are direct, retrograde, and in every conceivable direction.

It is remarked by a late writer, that the average *inclinations* of all the planes in which the comets now on record have been found to move, is about 90°. This he regards as a wonderful instance of the goodness of Providence, in causing their motions to be performed in a manner least likely to come in contact with the earth and the other planets.

307. Of the *physical nature* of comets, little is known. That they are, in general, very *light* and *vapory* bodies, is evident from the fact that stars have sometimes been seen even through their densest portions, and are generally visible through their tails, and from the little attractive influence they exert upon the planets in causing perturbations. While Jupiter and Saturn often *retard* and delay comets for months in their periodic revolutions, comets have not power, in turn, to *hasten* the time of the planets for a single hour; showing conclusively that the relative masses of the comets and planets are almost infinitely disproportionate.

Such is the extreme lightness or tenuity of cometary bodies, that in all probability the entire mass of the largest of them, if condensed to a solid substance, would not amount to more than a few hundred pounds. Sir Isaac Newton was of opinion, that if the tail of the largest comet was compressed within the space of a cubic inch, it would not then be as dense as atmospheric air! The comet of 1770 got entangled, by attraction, among the moons of Jupiter, on its way to the sun, and remained near them for *four months;* yet it did not sensibly affect Jupiter or his moons. In this way the *orbits* of comets are often entirely changed.

308. Comets were formerly regarded as harbingers of famine, pestilence, war, and other dire calamities. In one or two instances, they have excited serious apprehension that the day of judgment was at hand, and that they were the appointed messengers of Divine wrath, hasting apace to burn up the world. A little reflection, however, will show that all such fears are groundless. The same unerring hand that guides the ponderous planet

307. *Physical nature* of comets? What proofs of their light and vapory character? (What said of their probable mass? Opinion of Newton? What said of the comet of 1770? What effect on orbits?)

308. How comets formerly regarded? Why no fears of collision? (What estimate of "chances?")

in its way, directs also the majestic comet; and where infinite wisdom and almighty power direct, it is almost profane to talk of collision or accident.

<small>Even those who have calculated the "chances" of collision—as if chance had any thing to do among the solar bodies—have concluded the chances of collision are about as one to 281,000,000—*i. e.*, like the chance one would have in a lottery, where there were 281,000,000 black balls, and but one white one; and where the white ball must be produced at the first drawing to secure a prize.</small>

309. Were a collision actually to take place between a comet and the earth, it is not probable that the former would even penetrate our atmosphere, much less dash the world to pieces. Prof. Olmsted is of opinion that in such an event, not a particle of the comet would reach the earth—that the portions encountered by her would be arrested by the atmosphere, and probably inflamed; and that they would perhaps exhibit, on a more magnificent scale than was ever before observed, the phenomena of shooting stars or meteoric showers. The idea, therefore, that comets are dangerous visitants to our system, has more support from superstition than from reason or science.

<small>The air is to us what the waters are to fish. Some fish swim around in the deep, while others, like lobsters and oysters, keep on the bottom. So birds wing the air, while men and beasts are the "lobsters" that crawl around on the bottom. Now there is no more probability that a comet would pass through the atmosphere, and injure us upon the earth, than there is that a handful of *fog* or vapor thrown down upon the surface of the ocean, would pass through and kill the shell-fish at the bottom.</small>

310. After all that is supposed to be known respecting comets, it must be admitted that they are less understood than any other bodies belonging to our system. "What regions these bodies visit, when they pass beyond the limits of our view; upon what errands they come, when they again revisit the central parts of our system; what is the difference between their physical constitution and that of the sun and planets; and what important ends they are destined to accomplish in the economy of the universe, are inquiries which naturally arise in the mind, but which surpass the limited powers of the human understanding at present to determine."

<small>309. What probable effect in case of collision? Prof. Olmsted's opinion? (Remark respecting the air, fish, lobsters, &c. ?)
310. Are we as well acquainted with comets as with other bodies of our system? What inquiries suggested? How answered?</small>

CHAPTER X.

OF THE SUN.

311. Of all the celestial objects with which we are acquainted, none make so strong and universal an impression upon our globe as does the *sun*. He is the great center of the solar system—a vast and fiery orb, kindled by the Almighty on the morn of creation, to cheer the dark abyss, and to pour his radiance upon surrounding worlds. Compared with him, all the solar bodies are of inconsiderable dimensions; and without him, they would be wrapped in the gloom of interminable night.

312. The *form* of the sun is that of an oblate spheroid, his equatorial being somewhat greater than his polar diameter. The mean of the two is 886,000 miles. He is 1,400,000 times as large as the mighty globe we inhabit, and 500 times as large as all the planets put together. Were he placed where the earth is, he would fill all the orbit of the moon, and extend 200,000 miles beyond it in every direction. It would take 112 such worlds as ours, if laid side by side, to reach across his vast diameter.

THE SUN AND THE MOON'S ORBIT.

1. The vast magnitude of the sun may be inferred from the fact, that when rising or setting, he often *appears* larger than the largest building, or the tops of the largest trees. Now if the angle filled by him at the distance of two miles is over 100 feet across, what must it be at the distance of 95 millions of miles?

2. Were a railroad passed through the sun's center, and should a train of cars start from one side, and proceed on at the rate of 30 miles an hour, it would require 3¼ years

---

311. Describe the sun. How compare with the rest of the system?
312. What is his form? Diameter? Mass, as compared with our globe? With all other bodies of the system? With moon's orbit? (What sensible evidence of the vast magnitude of the sun? Illustration from railroad? Demonstration as to its comparison with moon's orbit?)

to cross over his diameter. To traverse his vast circumference, at the same rate of speed, would require nearly 11 years.

3. The mean distance of the moon from the earth's center is 240,000 miles; consequently the diameter of her orbit, which is twice the radius, is 480,000. Subtract this from 886,000, the sun's diameter, and we have 406,000 miles left, or 203,000 miles on each side, beyond the moon's orbit.

313. By the aid of telescopes, a variety of *spots* have been discovered upon the sun's disk. Their *number* is exceedingly variable at different times. From 1611 to 1629, a period of 18 years, the sun was never found clear of spots, except for a few days in December, 1624. At other times, twenty or thirty were frequently seen at once; and at one period, in 1825, upwards of fifty were to be seen. Prof. Olmsted states that over 100 are sometimes visible. From 1650 to 1670, a period of 20 years, scarcely any spots were visible; and for eight years, from 1676 to 1684, no spots whatever were to be seen.

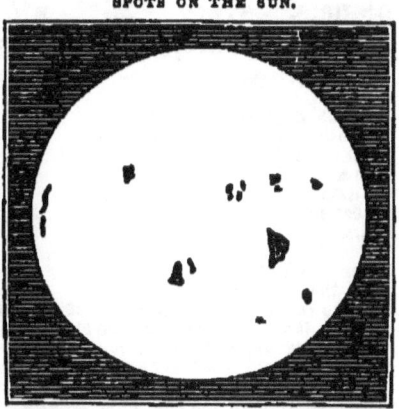

SPOTS ON THE SUN.

For the last 46 years, a greater or less number of spots have been visible every year. For several days, during the latter part of September, 1846, we could count sixteen of these spots, which were distinctly visible, and most of them well defined; but on the 7th of October following, only six small spots were visible, though the same telescope was used, and circumstances were equally favorable.

The sun is a difficult object to view through a telescope, even when the eye is protected in the best manner by colored glasses. In some cases (as in one related to the author by Professor Caswell, of Brown University), the heat becomes so great as to spoil the eye-pieces of the instrument, and sometimes the eye of the observer is irreparably injured.

314. The solar spots are all found within a zone 60° wide—*i. e.*, 30° each side of the sun's equator. They are generally permanent, though they have been known to

313. View of sun's surface through telescopes? Number of spots seen? Are they always to be seen? How from 1611 to 1629? In 1825? Prof. Olmsted's statement? How from 1650 to 1670? From 1676 to 1684? In 1846? (What said of difficulties of observing?)

314. Where are these spots situated? Are they permanent? What mo-

break in pieces, and disappear in a very short time. They sometimes break out again in the same places, or where none were perceptible before. They pass from left to right over the sun's disk in 13 days, 15 hours, and 45 minutes; from which it has been ascertained that he performs a sidereal revolution on his axis, from west to east, or in the direction of all the planets, every 25 days, 7 hours, and 48 minutes.

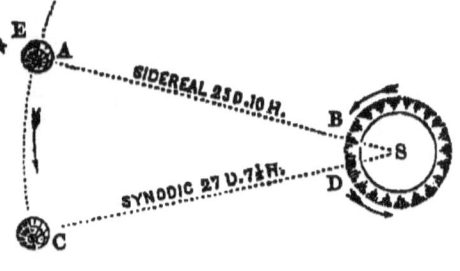

SIDEREAL AND SYNODIC REVOLUTIONS OF THE SUN.

1. His *apparent* or *synodic* revolution requires 27 days 7¼ hours; but this is as much more than a complete revolution upon his axis, as the earth has advanced in her orbit in 25 days 8 hours. Let S represent the sun, and A the earth in her orbit. When she is at A, a spot is seen upon the disk of the sun at B. The sun revolves in the direction of the arrows, and in 25 days 10 hours the spot comes round to B again, or opposite the star E. This is a *sidereal* revolution.

2. During these 25 days 8 hours, the earth has passed on in her orbit some 25°, or nearly to C, which will require nearly two days for the spot at B to get directly toward the earth, as shown at D. This last is a *synodic* revolution. It consists of one complete revolution of the sun upon his axis, and about 27° over.

315. Of the nature of these wonderful spots, a variety of opinions have prevailed, and many curious theories have been constructed. Lalande, as cited by Herschel, suggests that they are the tops of mountains on the sun's surface, laid bare by fluctuations in his luminous atmosphere; and that the penumbræ are the shoaling declivities of the mountains, where the luminous fluid is less deep. Another gentleman, of some astronomical knowledge, supposes that the tops of the solar mountains are exposed by *tides* in the sun's atmosphere, produced by planetary attraction.

To the theory of Lalande, Dr. Herschel objects that it is contradicted by the sharp termination of both the internal and external edges of the penumbræ; and advances as a more probable theory, that "they are the

---

tion have they? What conclusion from it? (What revolution is this? What time required for a synodic revolution? Illustrate.)

315. What are these spots supposed to be? Lalande? &c. Dr. Herschel's remark? Prof. Olmsted? Prof. Wilson? Experiments of Prof. Henry?

dark, or, at least, comparatively dark, solid body of the sun itself, laid bare to our view by those immense fluctuations in the luminous regions of the atmosphere, to which it appears to be subject." Prof. Olmsted supports this theory by demonstrating that the spots must be "nearly or quite in contact with the body of the sun."

In 1773, Prof. Wilson, of the University of Glasgow, ascertained, by a series of observations, that the spots were probably "*vast excavations* in the luminous matter of the sun;" the nuclei being their bottom, and the umbræ their shelving sides. This conclusion varies but little from that of Dr. Herschel, subsequently arrived at.

In a series of experiments conducted by Prof. Henry, of Princeton, by means of a thermo-electrical apparatus, applied to an image of the sun thrown on a screen from a dark room, it was found that the spots were perceptibly colder than the surrounding light surface.

316. The *magnitude* of the solar spots is as variable as their number. Upon this point, the second cut preceding gives a correct idea, as it is a pretty accurate representation of the sun's disk, as seen by the writer on the 22d of September, 1846. In 1799, Dr. Herschel observed a spot nearly 30,000 miles in breadth; and he further states, that others have been observed, whose diameter was upward of 45,000 miles. Dr. Dick observes that he has several times seen spots which were not less than $\frac{1}{25}$ of the sun's diameter, or 22,192 miles across.

It is stated, upon good authority, that solar spots have been seen by the naked eye—a fact from which Dr. Dick concludes that such spots could not be less than 50,000 miles in diameter. The observations of the writer, as above referred to, and represented in the cut, would go to confirm this deduction, and to assign a still greater magnitude to some of these curious and interesting phenomena.

317. The axis of the sun is inclined to the ecliptic $7\frac{1}{2}°$,

---

316. What said of the size of the solar spots? Dr. Herschel's observations? Dr. Dick's? The writer's?

or, more accurately, 7° 20'. This is but a slight deviation from what we may call a perpendicular; so that, in relation to the earth, he may be considered as standing up and revolving with one of his poles resting upon a point, just half his diameter below the ecliptic.

As the result of the sun's motion upon his axis, his spots always appear first on his eastern limb, and pass off or disappear on the west. But though the direction of the spots, as viewed from the earth, is from east to west, it only proves his motion to coincide with that of the earth, which we call from west to east; as when two spheres revolve in the *same* direction, the sides toward each other will appear to move in *opposite* directions. During one-half of the passage of the spots across the sun's disk, their apparent motion is *accelerated;* and during the remainder, it is *retarded*.

This apparent irregularity in the motion of the spots upon the sun's surface, is the necessary result of an equable motion upon the surface of a globe or sphere. When near the eastern limb, the spots are coming partly toward us, and their angular motion is but slight; but when near the center, their angular and real motions are equal. So, also, as the spots pass on to the west, it is their angular motion only that is diminished, while the motion of the sun upon his axis is perfectly uniform.

318. The figure of the sun affects not only the apparent *velocity* of the spots, but also their *forms*. When first seen on the east, they appear narrow and slender, as represented in the cut, page 147. As they advance westward, they continue to widen or enlarge till they reach the center, where they appear largest; when they again begin to contract, and are constantly diminished, till they disappear.

319. Another result of the revolution of the sun upon an axis inclined to the ecliptic, and the revolution of the

---

317. How is the sun's axis situated? What said of the direction of the spots? Of their rate of motion?
318. Of the *cause* of this irregularity? What variations in the *forms* of the solar spots? Cause?
319. What other result of the sun's revolution about an inclined axis? (Illustrate by diagrams.)

earth around him, is, that when viewed from our movable observatory, the earth, at different seasons of the year, the *direction* of the spots seems materially to vary

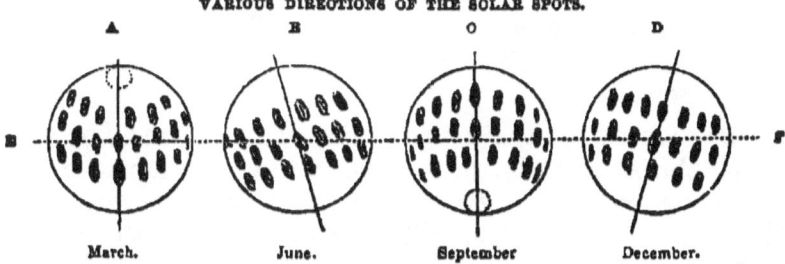

1. Let E F represent the plane of the ecliptic. In March, the spots describe a curve, which is convex to the south, as shown at A. In June, they cross the sun's disk in nearly straight lines, but incline upward. In September, they curve again, though in the opposite direction; and in December, pass over in straight lines, inclining *downward*. The figures B and D show the inclination of the sun's axis.

2. The *cause* of this difference in the direction of the solar spots will be fully understood by the following diagram:

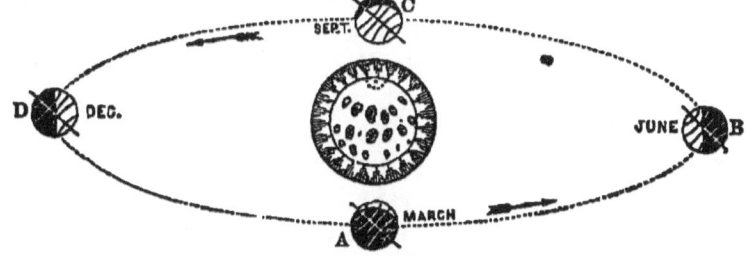

Let the student imagine himself stationed upon the earth at A, in March, looking upon the sun in the center, whose north or upper pole is now inclined *toward him*. The spots will then *curve downward*. Three months afterward—viz., in June—the earth will be at B; when the sun's axis will incline *to the left*, and the spots seem to pass *upward* to the right. In three months longer, the observer will be at C, when the north pole of the sun will incline *from him*, and the spots seem to *curve upward;* and in three months longer, he will be at D, when the axis of the sun will incline *to the right*, and the spots seem to incline downward.

320. Of the physical constitution of the sun, very little is known. When seen through a telescope, it is like a globe of fire, in a state of violent commotion or ebulition. La Place believed it to be in a state of actual combustion, the spots being immense caverns or craters, caused by eruptions or explosions of elastic fluids in the interior.

---

320. What said of the physical constitution of the sun? La Place's opinion? Most probable opinion?

The most probable opinion is, that the body of the sun is opake, like one of the planets; that it is surrounded by an atmosphere of considerable depth; and that the light is sent off from a luminous stratum of clouds, floating above or outside the atmosphere. This theory accords best with his density, and with the phenomena of the solar spots.

321. Of the *temperature* of the sun's surface, Dr. Herschel thinks that it must exceed that produced in furnaces, or even by chemical or galvanic processes. By the law governing the diffusion of light, he shows that a body at the sun's surface must receive 300,000 times the light and heat of our globe; and adds that a far less quantity of solar light is sufficient, when collected in the focus of a burning-glass, to dissipate gold and platina into vapor. The same writer observes that the most vivid flames disappear, and the most intensely ignited solids appear only as black spots on the disk of the sun, when held between him and the eye. From this circumstance he infers, that however dark the body of the sun may appear, when seen through its spots, it *may*, nevertheless, be in a state of most intense ignition. It does not, however, follow, of necessity, that it *must* be so. The contrary is, at least, physically possible. A *perfectly reflective* canopy would effectually defend it from the radiation of the luminous regions above its atmosphere, and no heat would be conducted downward through a gaseous medium increasing rapidly in density. The great mystery, however, is to conceive how so enormous a conflagration (if such it be) can be kept up from age to age. Every discovery in chemical science here leaves us completely at a loss, or rather seems to remove further from us the prospect of explanation. If conjecture might be hazarded, we should look rather to the known possibility of an indefinite generation of heat by friction, or to its excitement by the electric discharge, than to any actual combustion of ponderable fluid, whether solid or gaseous, for the origin of the solar radiation.

---

321. Sun's temperature? Dr. Herschel's idea? What reasoning against his opinion? What mystery?

322. The *Zodiacal Light* is a faint nebulous light, resembling the tail of a comet, or the milky way, which seems to be reflected from the regions about the sun, and is distinguishable from ordinary twilight. Its form is that of a pyramid or cone, with its base toward the sun, and inclined slightly to the ecliptic. It seems to surround the sun on all sides, though at various depths, as it may be seen in the morning preceding the sun, as well as in the evening following him; and the bases of the cones, where they meet at the sun, are much larger than his diameter.

ZODIACAL LIGHT.

323. Of the *nature* of this singular phenomenon, very little is positively known. It was formerly thought to be the atmosphere of the sun. Prof. Nichol says: "Of this, at least, we are certain—the zodiacal light is a phenomenon precisely similar in kind to the nebulous atmospheres of the distant stars, &c." Sir John Herschel remarks that it is manifestly of the nature of a thin lenticularly-formed atmosphere, surrounding the sun, and extending at least beyond the orbit of Mercury, and even of Venus. He gives the apparent angular distance of its vertex from the sun, at from 40° to 90°; and the breadth of its base, from 8° to 30°. It sometimes extends 50° westward, and 70° east of the sun at the same time.

324. The *form* of this substance surrounding the sun, and which is sufficiently dense to reflect his light to the

---

322. What is the zodiacal light? Its form? When seen?
323. Nature of this light? Former opinion? Prof. Nichol's remark? Dr. Herschel's? Its extent from the sun?
324. *Form* of this light? How situated with respect to sun's axis, &c.? (Illustrate by diagram.)

earth, seems to be that of a *lens;* or rather that of a huge wheel, thickest at the center, and thinned down to an edge at the outer extremities. Its being seen edgewise, and only one-half at a time, gives it the appearance of two pyramids with their bases joined at the sun. It is an interesting fact, stated by Prof. Nichol, that this light or nebulous body lies in the plane of the sun's equator. A line drawn through its transverse diameter, or from one apex of the pyramids to the other, would cross the axis of the sun at right angles. This fact would seem to indicate a revolution of this curious substance with the sun upon his axis.

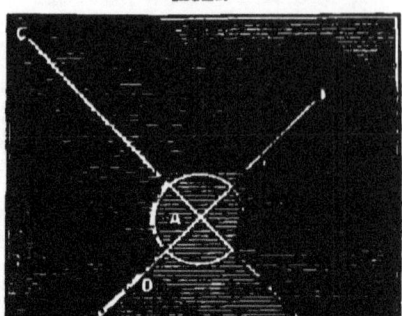

FORM, EXTENT, ETC., OF THE ZODIACAL LIGHT.

Let A, in the above cut, represent the sun, B B his axis; then C C will represent the extent, and D D the thickness of this curious appendage.

325. In regard to its atmospheric character, Dr. Dick observes that this opinion now appears extremely dubious; and Prof. Olmsted refers to La Place, as showing that the solar atmosphere could never reach so far from the sun as this light is seen to extend.

Another class of astronomers suppose this light, or rather the substance reflecting this light, to be some of the original matter of which the sun and planets were composed—a thin nebulous substance in a state of condensation, and destined either to be consolidated into new planetary worlds, during the lapse of coming ages, or to settle down upon the sun himself as a part of his legitimate substance. This theory will be noticed again when we come to speak of Nebulæ and Nebulous Stars.

326. After all the observations that have been made,

---

325. Remark of Dr. Dick respecting its atmospheric character? Olmsted and La Place? What other theory?

and the theories that have been advanced, it must be admitted that the subject of the zodiacal light is but imperfectly understood. Prof. Olmsted supposes it to be a nebulous body, or a thin vapory mass revolving around the sun; and that the *meteoric showers* which have occurred for several years in the month of November, may be derived from this body. This is the opinion of Arago, Biot, and others.

The best time for observing the zodiacal light is on clear evenings, in the months of March and April. It may be seen, however, in October, November, and December, before sunrise; and also in the evening sky.

THE SUN'S MOTION IN SPACE.

327. Although, in general terms, we speak of the sun as the *fixed center* of the system, it must not be understood that the sun is absolutely without motion. On the contrary, he has a periodical motion, in nearly a circular direction, around the common center of all the planetary bodies; never deviating from his position by more than twice his diameter. From the known laws of gravitation, it is certain that the sun is affected in some measure by the attraction of the planets, especially when many of them are found on the same side of the ecliptic at the same time; but this would by no means account for so great a periodical motion.

328. In addition to the motion above described, the sun is found to be moving, with all his retinue of planets and comets, in a *vast orbit*, around some distant and hitherto unknown center. This opinion was first advanced, we think, by Sir William Herschel; but the honor of actually determining this interesting fact belongs to Struve, who ascertained not only the *direction* of the sun and solar system, but also their *velocity*.

---

326. Is this subject well understood as yet? Prof. Olmsted's theory? When the best time for observing the zodiacal light?

327. Is the sun really stationary? What motion? How affected by planets?

328. What other motion? Who first advanced the opinion that he had such a motion? Who demonstrated it? Toward what point is the sun and

The point of tendency is toward the constellation Hercules, right ascension 259°, declination 35°. The *velocity* of the sun in space is estimated at 8 miles per second, or 28,000 miles per hour. Its period is about 18,200,000 years; and the arc of its orbit, over which the sun has traveled since the creation of the world, amounts to only about $\frac{1}{3000}$th part of his orbit, or about 7 minutes—an arc so small, compared with the whole, as to be hardly distinguishable from a straight line.

329. With this wonderful fact in view, we may no longer consider the sun as fixed and stationary, but rather as a vast and luminous *planet*, sustaining the same relation to some central orb that the primary planets sustain to him, or that the secondaries sustain to their primaries. Nor is it necessary that the stupendous mechanism of nature should be restricted even to these sublime proportions. The sun's central body may also have its orbit, and its center of attraction and motion, and so on, till, as Dr. Dick observes, we come to the great center of all —to the THRONE OF GOD!

Professor Mädler, of Dorpat, in Russia, has recently announced as a discovery that the star *Alcyone*, one of the seven stars, is the *center* around which the sun and solar system are revolving.

CHAPTER XI.

MISCELLANEOUS REMARKS UPON THE SOLAR SYSTEM.

NEBULAR THEORY OF THE ORIGIN OF THE SOLAR SYSTEM.

330. It was the opinion of La Place, a celebrated French astronomer, that the entire matter of the solar system, which is now mostly found in a consolidated

solar system tending? Its velocity? Period of revolution? Amount of its progress since the creation of the world?
329. How, then, should the sun be considered? How extend the analogy? What further recent discovery, and by whom?
330. State the " nebular theory" of the origin of the solar system? Who first started this theory?

state, in the sun and planets, was once a vast *nebula* or gaseous vapor, extending beyond the orbits of the most distant planets—that in the process of gradual condensation, by attraction, a *rotary motion* was engendered and imparted to the whole mass—that this motion caused the consolidating matter to assume the form of various concentric *rings*, like those of Saturn; and, finally, that these rings collapsing, at their respective distances, and still retaining their motion, were gathered up into *planets*, as they are now found to exist. This opinion is supposed to be favored, not only by the fact of Saturn's revolving rings, but by the existence of the zodiacal light, or a resisting medium about the sun; and also by the character of irresolvable or planetary nebulæ, hereafter to be described.

331. To this theory, however, there are many plausible, if not insurmountable, objections.

(*a*.) It seems to be directly at variance with the Mosaic account of the creation of the sun, moon, and stars. The idea that the sun and all the planets were *made up*, so to speak, out of the same general mass, not only throws the *creation* of this matter back indefinitely into eternity, but it substitutes the general law of attraction for the more direct agency of the Almighty. The creation spoken of in the Bible thus becomes not the *originating* of things that did not previously exist, but the mere *organization* or *arrangement* of matter already existing.

(*b*.) The supposed consolidation of the nebulous mass, in obedience to the general law of attraction, does not of itself account for the *rotary motion* which is an essential part of the theory. Under the influence of mere attraction, the particles must tend directly toward the center of the mass, and consequently could have no tendency to produce a rotary motion during the process of condensation.

(*c*.) The variation of the planetary orbits from the

---

331. What said of it? State the first objection named? The second? Third? Fourth? Fifth? What remark added by the author?

plane of the sun's equator contradicts the nebular theory. If the several primary planets were successively thrown off from the general mass, of which the sun is a part, they could not have been separated from the parent body till they were near the plane of its equator. Now, as the sun is assumed to be a part of the same mass, revolving still, the theory would require that the portions now separated from him, and called planets, should still revolve in the plane of his equator. But instead of this, it is found that some of them vary from this plane to the amount of near 42°.

(*d.*) This theory assumes not only that the primary planets were thrown off from the parent mass by its rapid revolution, but that the primaries, in turn, threw off their respective satellites. These, then, should all revolve in the plane of the planetary equators respectively, and in the direction in which their primaries revolve. But their orbits not only depart from the plane of the equators of their primaries (Jupiter's satellites excepted), but the moons of Uranus actually have a *retrograde* or *backward revolution*.

(*e.*) If the sun and planets are composed of what was originally the same mass, it will be necessary to show why they differ so materially in their physical natures— why the sun is self-luminous, and the planets opake.

But we have not room to discuss the subject at length in this treatise. It is but justice, however, to say, that men eminent for learning and piety have advocated the *nebular theory*, in the belief that it is perfectly consistent with the Mosaic account of creation. But the writer is frank to state, that while he acknowledges the force of some of the considerations urged in its support, he has not yet seen reason for adopting this theory of the origin of the solar system. "Through faith we understand that the worlds were framed by the word of God [not by the law of gravitation], so that things which are seen were not made of things which do appear [or of pre-existing matter]."—Heb. xi. 3.

332. Upon the supposition that the sun and planets were created as they are, by the direct act of God, an

inquiry at once arises as to the probable *extent* of the creation recorded by Moses. Does it include the whole universe? or is it to be understood as applicable only to the solar system? Upon this point our only light is, that "in the beginning God created the heavens and the earth"—that he not only made the sun and moon, but that "he made the stars also;" and that when these were spoken into being, God had "finished" his work. (See Genesis, 1st chapter.) "Thus the heavens and the earth were finished, and all the host of them." It seems most probable, therefore, that the Mosaic creation includes the whole material universe—that when God "laid the foundations of the earth," and the "heavens were the work of his hands," he "made the worlds also;" that is, they were *then* all "framed by the word of God."

WERE THE ASTEROIDS ORIGINALLY ONE PLANET?

333. Some very curious speculations have been entertained by astronomers in regard to the origin of the Asteroids. As in the case of the recently discovered planet, *Neptune*, the existence of a large planet between the orbit of Mars and Jupiter was *suspected* before the asteroids were known. This suspicion arose mainly from the seeming chasm that the absence of such a body would leave in the otherwise well-balanced solar system. The prediction that such a body would be discovered in the future stimulated the search of astronomers, till at length, instead of one *large* planet, *eighteen* small ones have, one after another, been discovered.

334. From certain peculiarities of the asteroids, it has been considered highly probable that they are the fragments of one large planet, which has been burst asunder by some great convulsion or collision. The grounds of this opinion are as follows:

---

332. What other interesting question started? What light upon this subject? What most probable?

333. What curious speculation respecting the asteroids? What suspicions before any of them were discovered?

334. What opinion respecting the origin of the asteroids? State the *grounds* of this opinion in order.

(*a.*) The asteroids are much *smaller* than any of the other primary planets.

(*b.*) They are all at nearly the same *distance* from the sun.

(*c.*) Their periodic revolutions are accomplished in nearly the same time. The difference of their periodic times is not greater than might result from the supposed disruption, as the parts thrown *forward* would have their motion *accelerated*, while the other parts would be thrown *back* or *retarded;* thus changing the periodic times of both.

(*d.*) The great departure of the *orbits* of the asteroids from the plane of the ecliptic is supposed to favor the hypothesis of their having been originally one planet, the assumption being that the explosion separating the original body into fragments would not only accelerate some portions and retard others, but would also throw them out of the plane of the original orbit, and in some cases still further from the ecliptic.

(*e.*) Their orbits are *more eccentric* than those of the other primaries. Although the tables show the eccentricity of Uranus's orbit as greater in miles than that of even Juno or Pallas, yet when we consider the difference in the *magnitude* of their orbits, it will easily be seen that his orbit is less elliptical than theirs.

(*f.*) The orbits of Ceres and Pallas, at least, *cross each other.* This, if we except, perhaps, the orbits of some of the comets, is a perfect anomaly in the solar system.

335. From all these circumstances, it has been concluded that the asteroids are only the fragments of an exploded world, which have assumed their present forms since the disruption, in obedience to the general laws of gravitation. This theory, first advanced by Dr. Olbers, is favored by Prof. Nichol, Dr. Brewster, Dr. Dick, and others; while Sir John Herschel observes that it may serve as a specimen of the *dreams* in which astronomers, like other speculators, occasionally and harmlessly in-

---

335. Who was the author of this theory? What distinguished astronomers favor it? What says Sir John Herschel? Remark of Dr. Dick? Opinion and remarks of the author?

dulge. Dr. Dick remarks that the breaking up of the exterior crust of the earth, at the time of the general deluge, was a catastrophe as tremendous and astonishing as the bursting asunder of a large planet. In view, however, of the *harmony* and *order* that everywhere reign throughout the planetary regions, directing the pathway and controlling the destiny of every world, it is hard to believe either that one world has been so constructed as to *explode*, like a vast bomb-shell, and scatter its fragments over the regions of its former pathway; or that He who guides even the erratic comet has allowed a ponderous world to get so off its track, as to dash itself to pieces against its fellow worlds.

ARE THE PLANETS INHABITED BY RATIONAL BEINGS?

336. Upon this interesting question, it must be admitted that we have no positive testimony. The argument for the inhabitedness of the planets rests wholly upon analogy, and the conclusion is to be regarded only in the light of a legitimate inference. Still, it is remarkable that those who are best acquainted with the facts of astronomy are most confident that other worlds as well as ours are the abodes of intellectual life. Indeed, as Dr. Dick well remarks, it requires a minute knowledge of the whole scenery and circumstances connected with the planetary system, before this truth comes home to the understanding with full conviction.

337. The analogies from which it is concluded that all the primary planets, at least, are inhabited by rational beings, are the following:

(*a.*) The planets are all *solid bodies* resembling the earth, and not mere clouds or vapors.

(*b.*) They all have a *spherical* or *spheroidal figure*, like our own planet.

(*c.*) The laws of gravitation, by which we are kept upon the surface of the earth, prevail upon all the other

---

336. What other question proposed? What admission? Nature of the evidence of the inhabitedness of the planets? What remarkable fact? Remark of Dr. Dick?

337. State the principal points of analogy between our globe and the other

14*

planets, as if to bind races of material beings to their surfaces, and provide for the erection of habitations and other conveniences of life. It is very remarkable, however, that those planets whose bulks are such as to indicate an insupportable attractive force, are not only less dense than our globe, but they have the most rapid daily revolution; as if, by diminished density, and a strong centrifugal force combined, to reduce the attractive force, and render locomotion possible upon their surfaces.

(*d.*) The *magnitudes* of the planets are such as to afford ample scope for the abodes of myriads of inhabitants. It is estimated that the solar bodies, exclusive of the comets, contain an area of 78,000,000,000 of square miles, or 397 times the surface of our globe. According to the population of England, this vast area would afford a residence to 21,875,000,000,000 of inhabitants; or 27,000 times the population of our globe.

(*e.*) The planets have a *diurnal revolution* around their axes, thus affording the agreeable vicissitudes of day and night. Not only are they opake bodies like our globe, receiving their light and heat from the sun, but they also revolve so as to distribute the light and shade alternately over each hemisphere. There, too, the glorious sun arises, to enlighten, warm, and cheer; and there "the sun-strown firmament" of the more distant heavens is rendered visible by the no less important blessing of a periodic night.

(*f.*) All the planets have an *annual revolution* round the sun; which, in connection with the inclination of their axes to their respective orbits, necessarily results in the production of *seasons*.

(*g.*) The planets, in all probability, are enveloped in *atmospheres*. That this is the case with many of them is certain; and the fact that a fixed star, or any other orb, is not rendered dim or distorted when it approaches their margin, is no evidence that the planets have no atmosphere. This appendage to the planets is known to vary in density; and in those cases where it is not de-

---

planets. Substance? Forms? Gravitation? Magnitude? Days and nights? Seasons? Atmospheres? Moons? Mountains? &c.

tected by its intercepting or refracting the light, it may be of a nature too clear and rare to produce such phenomena.

(*h.*) The principal primary planets are provided with *moons* or *satellites*, to afford them light in the absence of the sun. It is not improbable that both Mars and Venus have each, at least, one moon. The earth has one; and as the distances of the planets are increased, the number of moons seems to increase. The discovery of six around Uranus, and only one around Neptune, is no evidence that others do not exist which have not yet been discovered.

(*i.*) The surfaces of all the planets, primaries as well as secondaries, seem to be variegated with *hill and dale, mountain and plain*. These are the spots revealed by the telescope.

(*j.*) Every part of the globe we inhabit is destined to the support of animal life. It would, therefore, be contrary to the analogy of nature, as displayed to us, to suppose that the other planets are empty and barren wastes, utterly devoid of animated being. And if animals of any kind exist there, why not intelligent beings?

338. If other worlds are not the abodes of intellectual life, for what were they created? What influence do they exert upon our globe, especially those most remote? There are doubtless myriads of worlds beyond our system that will never even be seen by mortal eye, and that have no perceptible connection with our globe. If, then, they are barren and uninhabited islands in the great ocean of immensity, we repeat, for what were they created? The inquiry presses itself upon the mind with irresistible force, Why should this one small world be inhabited, and all the rest unoccupied? For what purpose were all these splendid and magnificent worlds fitted up, if not to be inhabited? Why these days and years—this light and shade—these atmospheres, and seasons, and satellites, and hill and dale?

---

338. What difficulty on the supposition that the planets are not inhabited?

339. To suppose all these worlds to be fitted up upon one general plan, provided with similar conveniences as abodes for intellectual beings, and yet only one of them to be inhabited, is like supposing a rich capitalist would build some thirty fine dwellings, all after one model, though of different materials, sizes, and colors, and provide in all for light, warmth, air, &c.; and yet, having placed the family of a son in *one* of them, allow the remaining twenty-nine to remain unoccupied forever! And as God is wiser than man, in the same proportion does it appear absurd, that of the twenty-six planetary temples now known to exist, only *one* has ever been occupied; while the remainder are mere specimens of Divine architecture, wheeling through the solitudes of immensity! The legitimate and almost inevitable conclusion, therefore, is, that our globe is only *one* of the *many worlds* which God has created to be inhabited, and which are now the abodes of his intelligent offspring. It seems irrational to suppose that we of earth are the only intelligent subjects of the "Great King," whose dominions border upon infinity. It is much more in keeping with sound reason, and with all the analogies of our globe, to suppose that

> "Each revolving sphere, a seeming point,
> Which through night's curtain sparkles on the eye,
> Sustains, like this our earth, its busy millions."

340. The fact that we neither *see*, nor *hear*, nor *hear from* the inhabitants of other worlds, is no evidence that such inhabitants do not exist. It would have been premature in Columbus had he concluded, when he saw land in the distance, that it was uninhabited, simply because he could not hear the shout of its savages, or see them gathered in groups upon the beach. So in regard to the distant planets. Our circumstances forbid our knowing positively that they are inhabited; so that the absence of that knowledge is no argument against the inhabitedness of other worlds.

---

339. What illustration? Conclusion? Poetry?
340. What said of the objection that we neither *see*, *hear*, nor *hear from* the inhabitants of the other worlds?

341. It may be thought that the extremes of heat and cold on some of the planets must be fatal to the idea of animal life, at least. But even this does not follow. Upon our globe, some animals live and flourish where others would soon die from heat or cold. And some animals, having cold blood, may be frozen, and yet live. So in other worlds. He who made the three Hebrews to live in the fiery furnace, can easily adapt the inhabitants of Mercury to their warm abode. And of the exterior planets we have only to say:

> " Who there inhabit must have other powers,
> Juices, and veins, and sense, and life, than ours;
> One moment's cold, like theirs, would pierce the bone,
> Freeze the heart's blood, and turn us all to stone!"

Adaptation is a law of the universe; and this at once obviates every difficulty in regard to the temperature of the planets, which might otherwise be urged as a reason why they were not inhabited.

---

341. Objection drawn from extremes of temperature? Poetry? What great law answers every such objection?

PART II.

THE SIDEREAL HEAVENS.

CHAPTER I.

THE FIXED STARS—CLASSIFICATION, NUMBER, DISTANCE, ETC.

342. THE sidereal heavens embrace all those celestial bodies that lie around and beyond the solar system, in the region of the fixed stars.

The fixed stars are distinguished from the planetary bodies by the following characteristics :

(*a.*) They shine *by their own light,* like the sun, and not by *reflection.*

(*b.*) To the naked eye, they seem to *twinkle* or *scintillate ;* while the planets appear tranquil and serene.

(*c.*) They maintain the same *general positions,* with respect to each other, from age to age. On this account, they are called *fixed stars.*

(*d.*) They are inconceivably *distant ;* so that, when viewed through a telescope, they present no sensible disk, but appear only as shining points on the dark concave of the sky. To these might be added several other peculiarities, which will be noticed hereafter.

343. For purposes of convenience, in finding or referring to particular stars, recourse is had to a variety of artificial methods of classification.

---

342. What parts of the book have we now gone over? Upon what do we now enter? What is meant by the sidereal heavens? How are the fixed stars distinguished from planetary bodies?

343. What are constellations? Their origin?

First, The whole concave of the heavens is divided into sections or groups of stars of greater or less extent. The ancients imagined that the stars were thrown together in clusters, resembling different objects, and they consequently named the different groups after the objects which they supposed them to resemble. These clusters, when thus marked out by the figure of some animal, person, or thing, and named accordingly, were called *constellations*.

344. Secondly, The stars are all classed according to their *magnitudes*. There are usually reckoned twelve different magnitudes, of which the first six only are visible to the naked eye, the rest being *telescopic stars*. These magnitudes, of course, relate only to their apparent brightness; as the faintest star may appear dim solely on account of its immeasurable distance. The method by which stars of different magnitudes are distinguished in astronomical charts is as follows:

STARS OF DIFFERENT MAGNITUDES.

"It must be observed," says Dr. Herschel, "that this classification into magnitudes is entirely arbitrary. Of a multitude of bright objects, differing, probably, intrinsically both in size and in splendor, and arranged at unequal distances from us, one must of necessity appear the brightest; the one next below it brighter still, and so on."

345. The next step is to classify the stars of *each constellation* according to their magnitude *in relation to each other*, and without reference to other constellations. In this classification, the Greek alphabet is first used. For instance, the largest star in Taurus would be marked (α) Alpha; the next largest (β) Beta; the next (γ) Gamma, &c. When the Greek alphabet is exhausted, the Roman or English is taken up; and when these are all absorbed, recourse is finally had to figures.

As Greek letters so frequently occur in catalogues and maps of the stars, and on the celestial globes, the Greek alphabet is here inserted, for the benefit of those who are not

---

344. How classified by *magnitudes?* (Remark of Dr. Herschel?)
345. Next step in classifying? How conducted? Greek letters? (Repeat the alphabet.)

acquainted with it; but as the capitals are seldom used for designating the stars, the small characters only are given:

THE GREEK ALPHABET.

α	Alpha	a	ν	Nu	n
β	Beta	b	ξ	Xi	x
γ	Gamma	g	ο	Omicron	o short
δ	Delta	d	π	Pi	p
ε	Epsilon	e short	ρ	Rho	r
ζ	Zeta	z	ς	Sigma	s
η	Eta	e long	τ	Tau	t
θ	Theta	th	υ	Upsilon	u
ι	Iota	i	φ	Phi	ph
κ	Kappa	k	χ	Chi	ch
λ	Lambda	l	ψ	Psi	ps
μ	Mu	m	ω	Omega	o long

346. To aid in finding particular stars, and especially in determining their numbers, and detecting changes, should any occur, astronomers have constructed *catalogues* of the stars, one of which is near 2,000 years old. Several of the principal stars have a specific name like the planets—as *Sirius, Aldebaran, Regulus*, &c.; and clusters of stars in a constellation sometimes receive a specific name, as the *Pleiades* and *Hyades* in Taurus.

347. The stars are still further distinguished into double, triple, and quadruple stars, binary system, variable stars, periodic stars, nebulous stars, &c.; all of which will be noticed hereafter.

NUMBER OF THE FIXED STARS.

348. The actual *number* of the stars is known only to Him who "telleth the number of the stars, and calleth them all by their names." The powers of the human mind are barely sufficient to form a vague estimate of the number near enough to be seen by our best telescopes, and here our inquiries must end.

The number of stars, down to the twelfth magnitude, has been estimated as follows:

---

346. What further methods for finding particular stars?
347. How are the stars still further distinguished?
348. Number of the stars? Of each magnitude? Number visible to naked eye? Additional seen through telescopes? Total? Remarks of Herschel and Olmsted?

Visible to the naked eye.		Visible only through telescopes.	
1st magnitude	18	7th magnitude	26,000
2d "	52	8th "	170,000
3d "	177	9th "	1,100,000
4th "	376	10th "	7,000,000
5th "	1,000	11th "	46,000,000
6th "	4,000	12th "	300,000,000
Total	5,623	Grand total,	354,301,623

Of these stars, Dr. Herschel remarks that from 15,000 to 20,000 of the first seven magnitudes are already *registered*, or noted down in catalogues; and Prof. Olmsted observes that Lalande has registered the positions of no less than 50,000.

349. The reason why there are so many more of the small stars than of the large ones is, that we are in the midst of a great cluster, with but few stars near us, the number increasing as the circumference of our view is enlarged. (See second cut, page 28, and also the adjoining.)

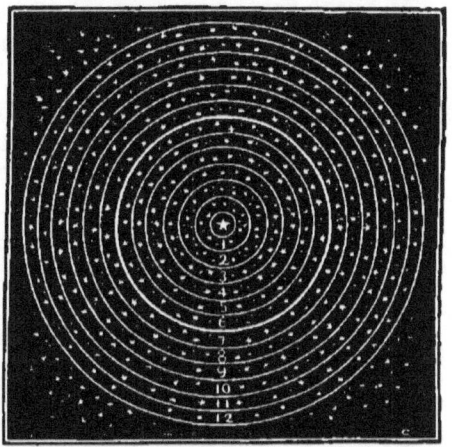

NUMBER OF STARS OF EACH MAGNITUDE.

Let the central star represent the sun (a star only among the rest), with the solar system revolving between him and the first circle. The 18 stars in space 1st will appear to be of the first magnitude, on account of their *nearness*, and they are thus few because they embrace but a small part of the entire cluster. The stars of space 2d will appear smaller, being more distant; but as it embraces more space, they will be more numerous. Thus as we advance from one circle to another, the apparent magnitude constantly diminishes, but the *number* constantly increases. The large white circle marks the limit of our natural vision. Even this cut fails to present fully to the eye the cause of the rapid increase in numbers, for we can only show the surface of a *cut section* of our firmament of stars, which exhibits the increase in a plane only; whereas our sun seems to be imbedded in the midst of a magnificent cluster (like a single apple in the midst of a large tree richly laden with fruit), the stars of which we view around us *in every direction.*

---

349. Why so many more of small stars than of the larger? (Illustrate by diagram. Does this convey a complete idea of the position of the sun, with reference to the fixed stars? Why not? What does his position more nearly resemble?)

350. If we suppose that each of these suns is accompanied only by as many planets as are embraced in our solar system, we have *nine thousand millions of worlds* in our firmament. No human mind can form a conception of this number; but even these, as will hereafter be shown, form but a minute and comparatively insignificant portion of the boundless empire which the Creator has reared, and over which he reigns. "Lo, these are parts of his ways; but how little a portion is heard of Him? but the thunder of his power who can understand." (Job xxvi. 14.)

DISTANCES AND MAGNITUDES OF THE STARS.

351. It has been demonstrated that the *nearest* of the fixed stars cannot be less than 20,000,000,000 (*twenty billions*) of miles distant! For light to travel over this space, at the rate of 200,000 miles per second, would require 100,000,000 seconds, or upwards of three years.

What, then, must be the distances of the telescopic stars, of the 10th and 12th magnitudes? "If we admit," says Dr. Herschel, "that the light of a star of each magnitude is half that of the magnitude next above it, it will follow that a star of the first magnitude will require to be removed to 362 times its distance, to appear no larger than one of the twelfth magnitude. It follows, therefore, that among the countless multitude of such stars, visible in telescopes, there must be many whose light has taken at least a thousand years to reach us; and that when we observe their places, and note their changes, we are, in fact, reading only their history of a thousand years' date, thus wonderfully recorded." Should such a star be struck out of existence now, its light would continue to stream upon us for a thousand years to come; and should a new star be created in those distant regions, a thousand years must pass away before its light could reach the solar system, to apprise us of its existence.

---

350. What supposition and conclusion? Scripture quotation?
351. Distances of the nearest stars? Time for light to travel over this space? Suppositions and conclusions of Dr. Herschel?

352. From what we have already said respecting the almost inconceivable distances of the fixed stars, it will readily be inferred that they must be bodies of great magnitude, in order to be visible to us upon the earth. It is probable, however, that "one star differeth from another" in its intrinsic splendor or "glory," although we are not to infer that a star is comparatively small because it appears small to us.

353. The prevailing opinion among astronomers is, that what we call the fixed stars are so many *suns* and centers of other systems. From a series of experiments upon the light received by us from Sirius, the nearest of the fixed stars, it is concluded that if the sun were removed 141,400 times his present distance from us, or to a point thirteen billions of miles distant, his light would be no stronger than that of Sirius; and as Sirius is more than twenty billions of miles distant, he must, in intrinsic magnitude and splendor, be equal to two suns like ours. Dr. Wollaston, as cited by Dr. Herschel, concludes that this star must be equal in intrinsic light to nearly fourteen suns. According to the measurements of Sir Wm. Herschel, the diameter of the star *Vega* in the Lyre is 38 times that of the sun, and its solid contents 54,872 times greater! The star numbered 61 in the Swan is estimated to be 200,000,000 miles in diameter.

354. Sir John Herschel states, that while making observations with his forty-feet reflector, a star of the first magnitude was unintentionally brought into the field of view. "Sirius," says he, "announced his approach like the dawn of day;" and so great was his splendor when thus viewed, and so strong was his light, that the great astronomer was actually driven from the eye-piece of his telescope by it, as if the sun himself had suddenly burst upon his view.

352. What inference from the great *distance* of the stars? What probability as to the real magnitude of the stars?
353. The prevailing opinion among astronomers? Conclusions from experiments with Sirius? Magnitude of *Vega?* Of No. 61 in the Swan?
354. Incident stated by Dr. Herschel? (Relative light of the stars of the first six magnitudes?)

According to Sir Wm. Herschel, the relative light of the stars of the first six magnitudes is as follows:

Light of a star of the average	1st magnitude				100
" " " " "	2d	"			25
" " " " "	3d	"			12
" " " " "	4th	"			6
" " " " "	5th	"			2
" " " " "	6th	"			1

CHAPTER II.

DESCRIPTION OF THE CONSTELLATIONS.

355. ALTHOUGH this work is designed particularly to illustrate the mechanism of the heavens, as displayed in the solar system, we are desirous of furnishing the learner with a sufficient guide to enable him to extend his inquiries and investigations not only to the different *classes* of bodies lying beyond the limits of the solar system in the far off heavens, but also, to the *constellations*, as such. For this purpose, we shall here furnish a brief description of the principal constellations visible in the United States, or in north latitude; by the aid of which, the student will be able to trace them, with very little difficulty, upon that glorious celestial atlas which the Almighty has spread out before us.

If the student will be at the trouble to identify the constellations by the aid of these descriptions, and without the aid of charts, it will give him a practical familiarity with the heavens which can be acquired in no other way. Indeed, this exercise is indispensable to a competent knowledge of sidereal astronomy, even where maps of the constellations are used. Let all students, therefore, embrace every favorable opportunity for looking up the constellations.

Those who wish to study their *mythological history* will consult the author's edition of the "*Geography of the Heavens*," by E. H. Burritt—the most reliable and popular work upon this subject in the English language.

356. Of the *nature* and *origin* of the constellations we have already spoken, at 343. Their formation has been the work of ages. Some of them were known at least 3,000 years ago. In the 9th chapter of Job, we

---

355. Principal design of this text-book? What further object? What done for this purpose? (Substance of note?)

356. What said of the *formation* of the constellations? Antiquity? Scripture allusions?

read of "Arcturus, Orion, and Pleiades, and the chambers of the south;" and in the 38th chapter of the same book, it is asked, "Canst thou bind the sweet influences of Pleiades, or loose the bands of Orion? Canst thou bring forth Mazzaroth in his season? or canst thou guide Arcturus with his sons?"

357. The constellations are distinguished into *ancient* and *modern*. According to Ptolemy's catalogue, the ancients had only 48 constellations; but being found convenient in the study of the heavens, new ones were added to the list, composed of stars not yet made up into hydras and dragons, till there is now scarcely stars or room enough left to construct the smallest new constellation, in all the spacious heavens. The present number, according to the catalogue of the Observatory Royal of Paris, is 93.

358. The constellations are further divided into the *Zodiacal*, *Northern*, and *Southern*. The *zodiacal* constellations are those which lie in the sun's apparent path, or along the line of the zodiac. The *northern* are those which are situated between the zodiacal and the north pole of the heavens; and the *southern*, those which lie between the zodiacal and the south pole of the heavens. They are distributed as follows—viz., 12 zodiacal, 35 northern, and 46 southern.

<small>This division is convenient for reference; but in tracing the constellations in the heavens, or upon a map, it is better to begin with those that are on or near the meridian, and proceed eastward, taking northern and southern together, so far as they are in view. And where classes in astronomy are organized during the fall months, it will be found advantageous to begin with the constellations that are in view at seasonable hours during those months.</small>

359. In consequence of the eastward motion of the earth in its annual revolution, the constellations rise earlier and earlier every night; so that if an observer were to watch the stars from the same position for a whole year, he would see each constellation, in turn, coming to the meridian at midnight (or at any other hour fixed

---

<small>357. How are the constellations classified? How many of each? In all?
358. How further classified? Describe each. How many of each? (What said in note?)
359. What said of the *rising* of the constellations? How proceed in describing and tracing?</small>

upon), till he had seen the whole panorama of the heavens. Beginning, therefore, with the constellations that are on or near the meridian at 9 o'clock, on the 15th of November, and going eastward, we shall now proceed with our description of the constellations.

OCTOBER, NOVEMBER, AND DECEMBER.

360. ANDROMEDA.—Almost directly over head, at 9 o'clock, on the 15th of November, may be seen the constellation *Andromeda*. The figure is that of a *woman* in a sitting posture, with her head to the southwest. Andromeda may be known by three stars of the second magnitude, situated about 12° apart, nearly in a straight line, and extending from east to west. The middle star of the three is situated in her *girdle*, and is called *Mirach*. The one west of Mirach is *Alpherat*, in the *head* of Andromeda; and the eastern one, called *Almaak*, is in her *left foot*. The star in her head is in the equinoctial colure. The three largest stars in this constellation are of the second magnitude. Near Mirach, are two stars of the third and fourth magnitudes, and the three in a row constitute the girdle.

This constellation embraces 66 stars, of which three are of the 2d magnitude, two of the 3d, and the rest small. About 2° from ν, at the northwestern extremity of the girdle, is a remarkable cluster or nebula of very minute stars, and the only one of the kind which is ever visible to the naked eye. It resembles two cones of light, joined at their base, about $\frac{3}{4}$° in length, and $\frac{1}{2}$° in breadth.

361. PEGASUS (*the Flying Horse*).—The figure is the head and fore parts of a *horse*, with wings. The three principal stars are of the 2d magnitude—viz., *Algenib*, about 15° south of Alpherat, in Andromeda; *Markab*, about 18° west of Algenib; and *Skeat*, 15° north of Markab. These three, with Alpherat in Andromeda, form what is called the *Square of Pegasus*. The head of the figure is to the southwest, almost in a line with Algenib and Markab, and about 20° from the latter.

---

360. Constellations on the meridian, in what months taken up? *Andromeda*—where situated? Figure? Position? How known? Name principal stars. (How many stars in constellation? What cluster, and where?)

361. Figure of Pegasus? Principal stars? How situated? Forming what? How the horse situated? His head where

362. PISCES (the Fishes) consists of two fishes, distinguished as the *northern* and *western*, connected by an irregular line of stars.

The *Western Fish* is situated directly south of the square of Pegasus—is about 20° long, with its head to the west. It includes a number of small stars, just south of Pegasus.

The *Northern Fish* is about the same size, with its head near Mirach in Andromeda, and its body extending to the north. This, also, includes small stars only, and is by no means conspicuous.

The *flexure* or *ribbon*, uniting the tails of the northern and western fishes, extends eastward from the latter, from star to star, till it comes opposite the former, when it turns to the north, taking several small stars in its way, till it joins the northern fish.

363. AQUARIUS (the Water-bearer) is represented by the figure of a man in a reclining posture, with his head to the northwest. Its four largest stars are of the third magnitude. It is situated directly south of the head of Pegasus, and from 5° to 30° north of a star of the first magnitude, in the southern fish. Three of the principal stars of Aquarius are near each other in the water-pot which he holds in his right hand.

364. PISCES AUSTRALIS (the Southern Fish) is situated directly south of Aquarius. Its largest star is *Fomalhaut*, of the 1st magnitude, which constitutes the *eye* of the fish. The body extends westward about 20°.

365. GRUS (the Crane) is situated directly south of the southern fish, with its head to the north. It is composed of a few stars only, of the fourth magnitude. As it is 45° south of the equinoctial, it appears low down in the south to persons situated in the Middle or Eastern States.

366. THE PHŒNIX is about 25° east of the Crane. It

---

362. Describe *Pisces*. The Western Fish? The Northern? Flexure?
363. Figure of *Aquarius?* Largest stars? Situation and extent? Further description.
364. *Pisces Australis*—largest star? Situation of figure?
365. *Grus*—how situated? Where? Composition?
366. Situation of the *Phœnix?* Principal stars?

has two stars of the 2d magnitude, about 12° apart east and west. The most western of these, in the neck of the bird, is about 25° southeast of Fomalhaut, in the Southern Fish. The other stars of the figure are of the 3d and 4th magnitudes.

367. CASSIOPEIA (the Queen).—About 30° northeast of Andromeda is Cassiopeia. The figure is that of a *woman* sitting in a *chair*, with her head from the pole, and her body in the Milky Way. Its four largest stars are of the 3d magnitude.

368. PERSEUS (the King).—Directly north of the "seven stars," and east of Andromeda, is Perseus. The figure is that of a man with a sword in his right hand, and the head of Medusa in his left. *Algol*, a star of the 2d magnitude, is about 18° from the Pleiades (or seven stars), in the head of Medusa; and 9° northeast of Algol is *Algenib*, of the same magnitude, in the back of Perseus. It embraces four other stars of the third magnitude, besides many smaller.

369. MUSCA (the Fly) is about 12° south of Medusa's head. It is a very small constellation, embracing one star of the 2d magnitude, two of the 3d, and a few smaller.

370. THE TRIANGLES include a few small stars, about half-way between Musca in the southeast, and Mirach in Andromeda in the northwest. Its two principal stars are of the 3d magnitude.

371. ARIES (the Ram).—The head of Aries is about 10° south of the Triangles. It may be known by two stars about 4° apart, of the 3d and 4th magnitudes. The most northeasterly of the two is the brightest, and is called *Arietis*. The back of the figure is to the north, and the body extends eastward almost to the Pleiades.

---

367. Where is *Cassiopeia?* Figure? Situation? Largest stars?
368. *Perseus*—figure? Two principal stars? Names? Situation? Magnitude?
369. Where is *Musca?* Size? Composition?
370. The *Triangles*—where? Principal stars?
371. Where is *Aries?* How known? Which of two principal stars brightest? Name? How figure situated? Extent?

372. **Cetus** (the Whale).—Directly southeast of Arietis, and about 25° distant, is *Menkar*, a star of the 2d magnitude, in the mouth of Cetus. This is the largest constellation in the heavens. It is situated below or south of Aries. It is represented with its head to the east, and extends 50° east and west, with an average breadth of 20°. The *head* of Cetus may be known by five remarkable stars, 4° and 5° apart, and so situated as to form a regular pentagon, or five-sided figure. About 40° southwest of Menkar, is another star in the body of the figure, near which are four small stars nearly in a row, and close together, running east and west.

Passing eastward, we next take the constellations that are on the meridian in

JANUARY, FEBRUARY, AND MARCH.

373. **Taurus** (the Bull) will be readily found by the seven stars or *Pleiades*, which lie in his neck. The largest star in Taurus is *Aldebaran*, in the Bull's eye, a star of the first magnitude, of a reddish color, somewhat resembling the planet Mars. Aldebaran, and four other stars in the face of Taurus, compose the *Hyades*. They are so placed as to form the letter V.

374. **Orion** lies southeast of Taurus, and is one of the most conspicuous and beautiful of the constellations. The figure is that of a man in the act of assaulting the bull, with a sword in his belt, and a club in his right hand. It contains two stars of the first magnitude, four of the second, three of the third, and fifteen of the fourth. *Betelguese* forms the right, and *Bellatrix* the left shoulder. A loose cluster of small stars forms the head. Three small stars, forming a straight line about 3° in length, constitute the *belt*, called by Job "the *bands of Orion*." They are sometimes called the *Three Kings*, because they point out the Hyades and Pleiades on the one hand,

---

372. *Cetus*—what star pointed out? Size of constellation? Situation? Extent? How know its *head?* What other star pointed out? What constellations next described in order?
373. *Taurus*—how found? Largest star? Hyades?
374. *Orion*—situation? Character? Figure? Composition?

and Sirius on the other. A row of very small stars runs down from the belt, forming the *sword*. These, with the stars of the belt, are sometimes called the *Ell and Yard*. *Mintika*, the northernmost star in the belt, is less than $\frac{1}{2}°$ south of the equinoctial. *Rigel*, a bright star of the first magnitude, is in the left foot, 15° south of Bellatrix; and *Saiph*, of the third magnitude, is situated in the right knee, $8\frac{1}{2}°$ east of Rigel.

375. LEPUS (the Hare) is directly south of and near Orion. It may be known by four stars of the third magnitude, in the form of an irregular square. *Zeta*, of the fourth magnitude, is the first star, situated in the back, and about 5° south of Saiph in Orion. About the same distance below Zeta are the four principal stars, in the legs and feet.

376. COLUMBA NOACHI (Noah's Dove) lies about 16° south of Lepus. It contains but four stars, of which *Phaet* is the brightest. It lies on the right, a little higher than *Beta*, the next brightest. This last may be known by a small star just east of it.

377. ERIDANUS (the River Po) is a large and irregular constellation, very difficult to trace. It is 130° in length, and is divided into the *northern* and *southern* streams. The former lies between Orion and Cetus, commencing near Rigel in the foot of Orion, and flowing out westerly in a serpentine course, near 40°, to the Whale.

378. CANIS MAJOR (the Greater Dog) lies southeast of Orion, and may be readily found by the brilliancy of its principal star, *Sirius*. This is the largest of the fixed stars, and is supposed to be the nearest to the solar system.

379. ARGO NAVIS (the Ship Argo) is a large and splendid constellation southeast of Sirius, but so low down in the south that but little of it can be seen in the United

---

375. Where is *Lepus?* How known? Describe.
376. *Columba Noachi*—situation? Composition?
377. Describe *Eridanus*. Length? Division? Situation?
378. Where is *Canis Major* situated? How found? What of *Sirius?*
379. Describe *Argo Navis*. Where situated? Principal stars, and where?

States. It lies southeast of Canis Major, and may be known by the stars in the prow of the ship. *Markeb*, of the fourth magnitude, is 16° southeast of Sirius. *Naos* and γ, still further south, are of the second magnitude, and *Canopus* and *Miaplacidus* of the first.

380. CANIS MINOR (the Lesser Dog) is situated about 25° northeast of Sirius, and between Canis Major and Cancer. It is a small constellation, having one star, *Procyon*, of the 1st magnitude, and *Gomelza*, of the 2d.

381. MONOCEROS (the Unicorn).—A little more than half way from Procyon to Betelguese in Orion, are three stars in a row, about 4° apart, and of the 4th magnitude. They extend from northeast to southwest, and constitute the face of Monoceros. His head is to the west, with Canis Minor on his back, and his hind feet about 25° southeast of Procyon. It is a large constellation, with but few stars, and those mostly small.

382. HYDRA (the Water Serpent).—About 20° east of Procyon are four stars of the fourth magnitude, situated about 4° apart, and so as to form a *diamond;* the longer axis running east and west. These constitute the *head* of *Hydra*, which points to the west. The figure extends to the south and east more than 100°, taking in an irregular line of stars of the 3d and 4th magnitudes. The largest star is about 15° southeast of the head. It is of the 2d magnitude, and is called *Alphard*.

383. CANCER (the Crab) is the least remarkable of the zodiacal constellations. It is situated about 15° north of the diamond in Hydra. It has no stars larger than the 3d magnitude, and is distinguished for a group of small stars called the Nebula of Cancer, which is often mistaken for a comet. A common telescope resolves this nebula into a beautiful assemblage of bright stars.

384. GEMINI (the Twins) may be known by two bright

---

380. *Canis Minor*—where? Describe.
381. Where is *Monoceros?* How situated? Composed? Character?
382. Where is the head of *Hydra?* How formed? Extent and position? Largest star?
383. Describe *Cancer.* Situation? Composition? For what distinguished?

stars of the 2d magnitude—one in the head of each figure. They are about 5° apart; the northeasterly one, and the brightest of the two, being about 25° due north of Procyon. This is *Pollux;* and the other one is called *Castor.* The *bodies* of the Twins extend from Castor and Pollux about 15° to the southwest, or toward Betelguese, in the right shoulder of Orion.

"This constellation," says Dr. Adam Clark, "was deemed propitious to mariners;" and on this account, the ship in which St. Paul sailed from Alexandria (Acts xxviii. 11) had the sign of Castor and Pollux.

385. HERSCHEL'S TELESCOPE covers two stars of the 5th magnitude, near each other, and about 10° north of Castor; and one other star of the same magnitude, about 10° northwest of the two first named. It is a small affair to immortalize Herschel's grand telescope.

386. THE LYNX is situated between Gemini and Cancer on the south, and the Pole in the north, the head being to the northwest. It has no stars larger than the 4th magnitude, and these are in two pairs—the first 15° northeast of Cancer, and the other 30° north of it. It is a loose and tame constellation, with nothing striking or peculiar by which it may be identified.

387. CAMELOPARDALIS (the Camelopard) extends from Perseus to the Pole. This, too, is a tame and uninteresting constellation, with but few stars in it, and those of the 4th magnitude, or less. The hind feet of the figure touch the Milky Way, and the head is composed of two stars of the 5th magnitude, 5° and 10° from the Pole star, toward the "dipper" in the Great Bear.

We now pass eastward to constellations that are on the meridian in

APRIL, MAY, AND JUNE.

388. URSA MAJOR (the Great Bear) is one of the most conspicuous in the northern heavens. It may be known

---

384. *Gemini*—how known? Names and situation of principal stars? Of figures? (Note.)
385. *Herschel's Telescope*—where? Character?
386. Situation of the *Lynx?* Position? Character?
387. Position of *Camelopardalis?* Extent? Character? Where the feet? The head, and how composed? What range of constellations next described?

by the figure of a large *dipper*, which constitutes the hinder part of the animal. This dipper is composed of seven stars. The first, in the end of the handle, is called *Benetnash*, and is of the 2d magnitude. The next is *Mizar*, known by a minute star almost touching it, called *Alcor*. *Mizar* is a double star. The third in the handle is *Alioth*. The first star in the bowl of the dipper, at the junction of the handle, is *Megrez*. Passing to the bottom of the dipper, we find *Phad* and *Merak*, while *Dubhe* forms the rim opposite the handle. Merak and Dubhe are called the *Pointers*, because they always point toward the Pole star. The *head* of the Great Bear lies far to the west of the Pointers (apparently *east* when seen *below* the Pole), and is composed of numerous small stars; while the *feet* are severally composed of two small stars, very near to each other. Megrez in Ursa Major, and Caph in Cassiopeia, are almost exactly opposite each other, on different sides of the Pole star, and about equally distant from it. They are both in the equinoctial colure.

389. Leo (the Lion).—About 55° southwest of the Pointers is *Regulus*, a star of the 1st magnitude, in the breast of Leo. This star is situated directly in the ecliptic. The *head* of the figure is to the west, the back being to the south. North of Regulus are several bright stars, in the form of a *sickle*, of which Regulus is the handle. *Denebola* is a bright star of the 2d magnitude, in the Lion's tail, about 25° northeast of Regulus, and 35° west of Arcturus.

390. Leo Minor (the Lesser Lion) is a small cluster of stars, of which one is of the 3d, and others of the 5th magnitude, about half way between Regulus and the Pointers. The head of the figure is northwest, and the principal stars form the body in the east, and the fore paws, which are extended to the west.

---

388. Describe *Ursa Major*. How known? Names of principal stars? Which are the *Pointers?* What said of *Megrez* and *Coph?*
389. Where is *Regulus?* In what constellation? How situated? Magnitude? How Leo placed? Where is the *sickle?* How constituted? Where is *Denebola?*
390. Describe *Leo Minor*. Where and how situated?

391. COMA BERENICES (Berenices' Hair) is a beautiful cluster of small stars, about 20° northeast of Denabola, and half way from Leo Minor to Arcturus. It has but one star as large as the 4th magnitude.

392. COR COROLI (Charles's Heart) is a bright star of the 3d magnitude, about 12° north of Coma Berenices. The figure includes several other stars, east and west, of the 5th magnitude.

393. BOOTES (the Bear-driver) is directly east of Coma Berenices. The figure is that of a man, with his head toward the Pole, and *Arcturus*, a star of the 1st magnitude, in the left knee. The other stars are of the 3d and 4th magnitudes. Three small stars, forming a triangle, and situated 15° northeast of Arcturus, mark the right hand of the figure ; while two stars of the 3d and 4th magnitudes, and still further north, mark his shoulders. The head is marked by *Nekkar*, another star of the 3d magnitude.

394. VIRGO (the Virgin) lies directly south of Coma Berenices and Bootes. The figure is that of a woman with wings, with her head to the west, near Denabola in Leo; and her feet about 40° to the east. *Spica*, the principal star, is of the 1st magnitude, about 35° southwest of Arcturus.

395. CRATER (the Cup) is composed of six small stars 30° west of Spica. The largest is of the 4th magnitude.

396. CORVIS (the Crow) is still nearer, being only 15° southwest of Spica. It has two stars of the 3d magnitude, and three of the 4th.

397. LIBRA (the Balance) is about 25° east of Spica. It has two stars of the 2d magnitude, about 10° apart, which, with two others of the 3d magnitude southeast of

---

391. *Coma Berenices*—character ? Situation ?
392. *Cor Coroli*—principal star ? Situation ?
393. Where is *Bootes ?* Figure ? Position ? Principal stars ?
394. Where is *Virgo* situated ? Figure ? Position ? Principal stars ?
395. *Crater*—how situated ? Largest star ?
396. *Corvis*—where ? Composition ?
397. Where is *Libra ?* Composition ?

them, form a small quadrilateral figure. Its few remaining stars are at the east, and of the 4th magnitude.

398. CENTAURUS (the Centaur) is a fine compact constellation about 30° south or southeast of Spica. It has nine stars of the 3d magnitude, mostly in the head of the figure. It is too low in the south to be visible in the United States, except when near the meridian.

JULY, AUGUST, AND SEPTEMBER.

399. URSA MINOR (the Lesser Bear) is composed of a few stars near the north pole of the heavens, and mostly of the 3d and 4th magnitudes. The back of the figure is toward the pole, with its head to the west. The *Pole star*, of the 2d magnitude, is in the extremity of its tail.

400. DRACO (the Dragon) is an irregular serpentine constellation, embracing a large circuit in the polar regions. He winds round between the Great and Little Bear, and, commencing with the tail, between the Pointers and Pole star, is easily traced, by a succession of bright stars extending from west to east. Passing south of Ursa Minor, around nearly to Cepheus, it returns westward, and terminates in four stars, which form the head, near the foot of Hercules. These four stars are 3°, 4°, and 5° apart, so situated as to form an irregular square; the two upper ones, *Etamis* and *Rastaban*, being the brightest, and both of the 2d magnitude.

401. HERCULES (the Giant) is a large, but not very striking or conspicuous constellation. The figure is that of a giant, with a large club in his right hand, and a hydra in his left. The head of the figure is to the south, and the whole is composed of stars from the 2d to the 4th magnitude. This constellation is thickly set with stars, the largest of which is called *Rasalgethi*, in the head

---

398. Describe *Centaurus*. Position? Composition?
399. *Ursa Minor*—position? Principal star?
400. *Draco*—position? How traced? Where head? How composed? Form what?
401. *Hercules*—figure? Situation? Composition? Principal star? Number of stars?

of the figure, and is of the 2d magnitude. It has nine stars of the 3d magnitude, and 19 of the 4th.

402. CORONI BOREALIS (the Northern Crown) is about 15° west of the middle of Hercules. Its principal star is *Alphacca*, a bright star of the 2d magnitude, about 20° northeast of Arcturus. About the same distance, directly east of Arcturus, is a small group of stars, which constitute the head of the *Serpent*.

403. SCORPIO (the Scorpion) is one of the most interesting and splendid of the constellations. It is situated about 45° east of Spica, adjoining Libra. The head of the figure is composed of five stars—one of the 2d, and the others of the 3d magnitude—forming an arc of a circle convex to the west. The largest of these five stars is in the ecliptic, and is called *Graffias*. About 9° southeast of Graffias is *Antares*, a star of the 1st magnitude, in the body of the figure, and of a reddish color. A number of bright stars of the 4th magnitude extend to the southeast into the Milky Way, and then curve around to the east and north, forming the tail of Scorpio.

404. LEPUS (the Wolf) consists of a small group of stars, about 15° southwest of Antares. The head of the figure is to the north.

405. SERPENTARIUS (the Serpent-bearer) is a large but uninteresting constellation, between Scorpio on the south, and Hercules on the north. The figure is that of a man grasping a serpent, the head of which has already been described (402). The *folds* of the serpent may be traced by a succession of bright stars extending for some distance to the east. The principal star in Serpentarius is of the 2d magnitude, and is called *Ras Alhague*. It is situated in the *head* of the figure, and within 5° of Rasalgethi, in the head of Hercules. The *feet* of the figure

---

402. *Coroni Borealis*—location? Principal star? What other group of stars mentioned?
403. Describe *Scorpio*. Situation? Composition? Largest star in head? What other large star? Position and composition of tail?
404. *Lepus*—composition? Position?
405. *Serpentarius*—situation? Figure? Principal star? Situation?

rest upon Scorpio, and the right shoulder touches the Milky Way.

406. LYRA (the Harp) is a small constellation 15° east of Hercules. Its principal star is *Vega*, of the 1st magnitude, one of the brightest stars in the northern hemisphere. It has two stars of the 3d magnitude, and several others of the 4th.

407. CYGNUS (the Swan) is situated directly east of Lyra. Three bright stars, which lie along the Milky Way, form the *body* and *neck* of the Swan, running northeast and southwest; and two others, at right angles, in a line with the middle one of the three, constitute the *wings*. These five stars form a large *cross*. *Arided*, in the body of the Swan, is a star of the 1st magnitude, and the remaining ones of the constellation are of the 3d and 4th.

408. THE FOX AND GOOSE is located just south of Cygnus, with the head to the west. It is a small constellation; the two principal stars of which, of the 2d magnitude, form the head of the Fox. Most of the figure is in the Milky Way.

409. AQUILA (the Eagle) is still south of Cygnus and the Fox. It is conspicuous for three bright stars in its neck, of which the central one, *Altair*, is a brilliant white star of the first magnitude, just east of the Galaxy.

410. DELPHINUS (the Dolphin) is a beautiful little cluster of stars, 15° northeast of the Eagle. It may be known by four principal stars in the head, of the 3d magnitude, arranged in the figure of a diamond, and pointing northeast and southwest. A star of the same magnitude, about 5° south, makes the tail.

411. ANTONIUS lies directly south of Aquila, his head being near Altair, and the body and feet to the southwest. Two stars of the 3d magnitude constitute the right

---
406. *Lyra*—situation? Principal star? What others?
407. *Cygnus*—situation? Composition?
408. *Fox and Goose*—location? Position of figure?
409. *Aquila*—where? For what conspicuous?
410. Describe *Delphinus*. How known?
411. *Antonius*—situation? How placed? Composition?

arm, and several smaller ones make the bow and arrows held in his hand.

412. SAGITTARIUS (the Archer) lies next to Scorpio, and may be known by three stars in the Galaxy, arranged in a curve, to represent the *bow* of the archer. The central star is the brightest, and has a bright star directly west of it, forming the head of the *arrow*. The head and chest of Sagittarius are just east of the Milky Way, between the tail of Scorpio and the head of Capricornus.

413. CAPRICORNUS (the Goat) is situated about 20° northeast of Sagittarius. The head of the figure is to the west, and is composed of two bright stars, of the 3d magnitude, and about 4° apart. There is a smaller star between them, and several still smaller close around them.

414. *CRUX (the Cross) is a brilliant little constellation, but too far south to be visible to us at the north. It consists of four principal stars—namely, one of the 1st, two of the 2d, and one of the 3d magnitude.

<small>Besides these, there are several fine constellations about the south pole of the heavens, as the *Altar*, the *Peacock*, *Charles's Oak*, &c.; but as they cannot be traced from the latitudes in which this book will be used, it is thought not important to describe them.</small>

---

412. *Sagittarius*—where? How known?
413. *Capricornus*—where? Position of figure? Composition?
414. *Crux*—describe. Composition? (What said of south circumpolar constellations? Names? Why not described?)

CHAPTER III.

DOUBLE, VARIABLE, AND TEMPORARY STARS, BINARY SYSTEMS, ETC.

415. MANY of the stars which, to the naked eye, appear single, are found, when examined by the aid of a telescope, to consist of two or more stars, in a state of near proximity to each other. These are called *Double Stars*. When three or more stars are found thus closely connected, they are called *Triple* or *Multiple Stars*.

416. Double and triple stars are supposed to be constituted in two ways—first, by actual contiguity; and secondly, where they are only near the same line of vision, one of the component stars being far beyond the other. In the former case, they are said to be *physically* double, from the belief that they are bound together by attraction, and that one revolves around the other; while in the latter case, they are considered as only *optically* double.

STARS OPTICALLY DOUBLE.

Apparent positions.   True positions.

A   B

Here the observer on the left sees a large and small star at A, apparently near together—the lowest star being much the smallest. But instead of their being situated as they appear to be, with respect to each other, the true position of the smaller star may be at B instead of A; and the difference in their apparent magnitudes may be wholly owing to the greater distance of the lower star.

Upon this subject Dr. Herschel remarks, that this nearness of the stars to each other, in certain cases, might be attributed to some accidental cause, did it occur only in a few instances; but the frequency of this companionship, the extreme closeness, and, in many cases, the near equality of the stars so conjoined, would alone lead to a strong suspicion of a more near and intimate relation than mere casual juxtaposition.

---

415. What said of double, triple, and multiple stars?
416. How are they supposed to be constituted? How distinguished? (Illustrate by diagram. Remark of Dr. Herschel? How many specimens of double stars given?)

The following will convey to the student an idea of the telescopic appearance of some of the double stars:

SPECIMENS OF DOUBLE STARS.

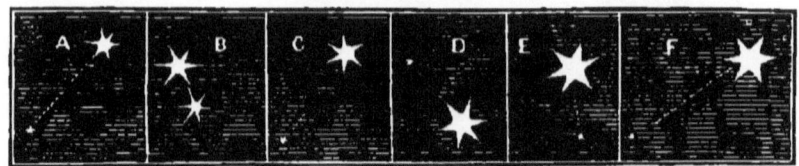

417. A is a double star in *Ursa Minor*, commonly known as the Pole star. It consists of a star of the 2d, and another of the 9th magnitude, situated about 18″ apart, or about four times the diameter of the larger star. They are both of a silvery white. It requires a pretty good telescope to show this star double; hence it is considered a very good *test object* by which to ascertain the qualities of a telescope, especially of the low-priced refractors.

The writer has often seen the companion of the Pole star distinctly, with a six-inch refracting telescope, manufactured by Mr. Henry Fitz, New York.

418. B is a view of the double star *Castor*, in the Twins. The stars are of a greenish white, of the 3d and 4th magnitudes, and about 5″, or two diameters of the principal star, apart. This also is considered a good test object. Through ordinary telescopes, the stars seem to be in contact; but with those of higher power, they appear fairly divided. These stars constitute what is called a *Binary System*.

419. C is a representation of *Mizar*, the middle star, in the tail of the Great Bear. It may be seen double with a good spy-glass. The stars are both of a greenish white, of the 3d and 4th magnitudes, and about 14′ apart. Mizar has sometimes been seen *without a companion*, and at other times it has been known suddenly to appear. The companion is not *Alcor*, near Mizar, and visible to the naked eye, but a *telescopic star*.

---

417. What is Fig. A in the cut? How composed? Color? How seen? (Remark of author in note?)
418. Fig. B—color? Magnitudes? Distance apart? Further remark?
419. Fig. C—how seen? Color? Magnitude? Distance? Additional remarks?

420. D is a view of the double star *Mintaka*, the middle star of the three forming the belt of Orion. The component stars are of the 4th and 8th magnitudes—the largest of a reddish hue, and the small one white. They are about 10″ apart, or four times the diameter of the largest star.

421. E is a view of *Rigel*, in the left foot of Orion. The components are of the 1st and 9th magnitudes, and about 10″ apart. Their color is a yellowish white.

422. F is a view of the bright star *Vega*, in the Lyre. Its companion is a star of the 11th magnitude, situated about 40″ distant. This is a close test object for an ordinary telescope.

423. The *number* of double stars has been variously estimated. Sir William Herschel enumerates upwards of 500, the individuals of which are within 30″ of each other. Professor Struve of Dorpat estimated the number at about 3,000; and more recent observations fix the number at not less than 6,000. The great number of the double stars first led astronomers to suspect a physical connection by the laws of gravitation, and also a *revolution* of star around star, as the planets revolve around the sun.

BINARY AND OTHER SYSTEMS.

424. By carefully noting the relative distances and angular positions of double and multiple stars, for a series of years, it has been found that many of them have their periodic revolutions around each other. Where two stars are found in a state of revolution about a common center, they constitute what is called a *Binary System*. These, it must be remembered, are the *double* and *multiple* stars, which appear single to the naked eye. Sir W. Herschel noticed about 50 instances of changes in the angular position of double stars; and the revolu

---

420. Fig. D—describe. Magnitude? Color? Distance?
421. Fig. E—place? Components? Distance? Color?
422. Fig. F—companion?
423. Number of double stars? Led to what?
424. Motions of double stars? What are *binary systems?*

tion of some *eighteen* of these is considered certain. Their periods vary from 40 to 1,200 years.

425. The star numbered 70 in the Serpent-bearer is a binary system. The periodic time of the revolving star is about 93 years. In the course of its revolution, the two stars sometimes appear separated, sometimes very near together, and at other times as one star. They are of the 5th and 6th magnitudes, and of a yellowish hue.

426. The star ξ, in the left hind paw of *Ursa Major*, is one of these stellar systems. The revolution of its component stars began to be noticed in 1781; since which time they have made one complete revolution, and are now (1853) some fourteen years on the second. Of course, then, their periodic time is about 58 years. Their angular motion is about 6° 24' per year.

Dr. Dick supposes these stars to be some 200,000,000,000 miles apart; and upon the supposition that the smaller revolves around the latter, computes its velocity to be not less than 2,471,000 miles every hour. This would be 85 times the velocity of Jupiter and 23 times the velocity of Mercury—the swiftest planet in the solar system.

427. The star γ in Virgo is another of these systems. It has been known as a double star for at least 130 years. The two stars are both of the 3d magnitude, and of a yellowish color. The late E. P. Mason, of Yale College, estimated its period at 171 years. More recent observations and estimates by Mädler give a period of 145 years.

428. "To some minds, not accustomed to deep reflection," says Dr. Dick, " it may appear a very trivial fact to behold a small and scarcely distinguishable point of light immediately adjacent to a larger star, and to be informed that this lucid point revolves around its larger attendant; but this phenomenon, minute and trivial as it may at first sight appear, proclaims the astonishing fact, that *suns revolve around suns, and systems around systems.* This is a comparatively new idea, derived from our late sidereal investigations, and forms one of the

425. Describe 70 Ophiuchi?
426. What specimen described? Period? Yearly angular motion? (Dr. Dick's remark?)
427. What other binary system? How long known? Components? Period?
428. Quotation from Dr. Dick.

most sublime conceptions which the modern discoveries of astronomy have imparted.

429. "It undoubtedly conveys a very sublime idea, to contemplate such a globe as the planet Jupiter—a body thirteen hundred times larger than the earth—revolving around the sun, at the rate of twenty-nine thousand miles every hour; and the planet Saturn, with its rings and moons, revolving in a similar manner round this central orb, in an orbit five thousand six hundred and ninety millions of miles in circumference. But how much more august and overpowering the conception of a sun revolving around another sun—of a sun encircled with a retinue of huge planetary bodies, all in rapid motion, revolving round a distant sun, over a circumference a hundred times larger than what has been now stated, and with a velocity perhaps a hundred times greater than that of either Jupiter or Saturn, and carrying all its planets, satellites, comets, or other globes, along with it in its swift career! Such a sun, too, may as far exceed these planets in size as our sun transcends in magnitude either this earth or the planet Venus; the bulk of any one of which scarcely amounts to the thirteen-hundred-thousandth part of the solar orb which enlightens our day.

430. "The further we advance in our explorations of the distant regions of space, and the more minute and specific our investigations are, the more august and astonishing are the scenes which open to our view, and the more elevated do our conceptions become of the grandeur of that Almighty Being who 'marshalled all the starry hosts,' and of the *multiplicity* and *variety* of arrangements he has introduced into his vast creation. And this consideration ought to serve as an argument to every rational being, both in a scientific and a religious point of view, to stimulate him to a study of the operations of the Most High, who is 'wonderful in counsel, and excellent in working,' and whose works in every

---

429. What further remarks?
430. Continue quotation. (What table? Note?)

part of his dominions adumbrate the glory of his perfections, and proclaim the depths of his wisdom, and the greatness of his power."

The following table shows the periodic times of the principal binary systems, so far as known:

BINARY SYSTEMS.

Names.	Period in years.	Names.	Period in years.
η Coronæ....	43·40	ω Leonis.....	82·533
ζ Cancri....	55·00	ξ Bootes.....	117·140
ξ Ursæ Majoris	58·26	a Hercules...	31·468
70 Ophiuchi...	93·00	b Ursæ Majoris	58·262
61 Cygni.....	452·00	c " "	61·464
γ Virginis...	145·00	p Ophiuchi...	73·862
Castor......	286·00	b " ...	80·340
σ Coronæ....	145·00	c " ...	92·870
γ Leonis.....	1200·00	β Coronæ....	608·450

The student should here be reminded that these are not systems of planets revolving around suns, but of *sun revolving around sun;* and that their component stars may not only be as far apart as our sun and Sirius, but that they are probably each the center of his own planetary system, like that which revolves around our central orb.

431. Besides the revolutions of these double stars around each other, they are found to have a proper motion together in space, like that which our sun has around the great central Sun. Upon this subject Sir John Herschel observes, that these stars not only revolve around each other, or about their common center of gravity, but that they are also transferred, without parting company, by a progressive motion common to both, toward some determinate region.

The two stars of 61 Cygni, which are nearly equal, have remained constantly at the same, or very nearly the same, distance of 15", for at least 50 years past. Meanwhile, they have shifted their local situation in the heavens, in this interval of time, through no less than 4' 23"—the annual proper motion of each star being 5".3; by which quantity (exceeding a third of their interval) this system is every year carried bodily along in some unknown path, by a motion which, for many centuries, must be regarded as uniform and rectilinear. Among stars not double, and no way differing from the rest in any other obvious particular, μ Cassiopeiæ is to be remarked as having the greatest proper motion of any yet ascertained, amounting to 8".74 of annual displacement.

431. What other motion of the stars? Dr. Herschel? (Specimen in note? Motions? What star named as having the greatest proper motion of any yet known?)

432. But though motions which require whole centuries to accumulate before they produce changes of arrangement, such as the naked eye can detect, are quite sufficient to destroy that idea of mathematical fixity which precludes speculation, yet are they too trifling, so far as practical applications go, to induce a change oᴀ language, and lead us to speak of the stars, in common parlance, as otherwise than fixed.

433. Most of the double, triple, and multiple stars are of *various colors*, beautifully contrasting with each other.

> "———————— Other suns, perhaps,
> With their attendant moons ————————,
> Communicating male and female light,
> (Which two great sexes animate the world,)
> Stored in each orb, perhaps, with some that live."

It is probable, however, that, in most cases, this variety of colors is merely *complimentary*, in accordance with that general law of optics which provides that when the retina is under the influence of excitement, by any bright colored lights, feebler lights, which, seen alone, would produce no sensation but of whiteness, shall for the time appear colored with the tint complimentary to that of the brighter. Thus, if a yellow color predominate in the light of the brighter star, that of the less bright one in the same field of view will appear blue; while, if the tint of the brighter star verge to crimson, that of the other will exhibit a tendency to green, or even appear as a vivid green, under favorable circumstances.

434. This first law of contrast is beautifully exhibited by ι Cancri—the latter by γ Andromedæ; both fine double stars. If, however, the colored star be much the less bright of the two, it will not materially affect the other. Thus, for instance, η Cassiopeiæ exhibits the beautiful combination of a large white star, and a small one of a rich ruddy purple.

It is by no means, however, intended to say, that in all such cases one of the colors is a mere effect of contrast; and it may be easier suggested in words than conceived in

432. Why called "fixed stars," if in motion?
433. What said of the color of double stars? Quotation from Milton? Cause of this variety of colors?
434. Specimens of complimentary colors? (Are they all complimentary?)

imagination what variety of illumination *two suns*—a red and a green, or a yellow and a blue one—must afford a planet circulating about either, and what charming contrasts and "grateful vicissitudes"—a red and a green day, for instance, alternating with a white one and with darkness—might arise from the presence or absence of one or other, or both, above the horizon. Insulated stars of a red color, almost as deep as that of blood, occur in many parts of the heavens; but no green or blue star, of any decided hue, has, we believe, aver been noticed unassociated with a companion brighter than itself.

VARIABLE OR PERIODICAL STARS.

435. Variable stars are those which undergo a regular periodical increase and diminution of lustre, amounting, in some cases, to a complete extinction and revival. These variations of brilliancy, to which some of the fixed stars are subject, are reckoned among the most remarkable of celestial phenomena. Some of them pass through their successive changes with great rapidity; while in other cases, their brilliancy is increased or diminished gradually for months. The time occupied by one of these stars, in passing through all their different phases, is called its *period*.

436. One of the most remarkable of these variable stars is the star *Omicron*, or *Mira* in the *Whale*. Its period is about 332 days, during which time it varies from a star of the 2d magnitude to complete invisibility. It appears about twelve times in eleven years—remains at its greatest brightness about a fortnight; being then, on some occasions, equal to a large star of the 2d magnitude. It then decreases for about three months, when it disappears. In about five months, it becomes visible again, and continues to increase during the remaining three months of its period.

Its increase of light is much more rapid than its decrease. It increases from the 6th to the 2d magnitude in 40 days, continues thus brilliant 26 days, and then fades to the 6th magnitude again in 66 days. Hence it is above the 6th magnitude for 132 days, and below 200 days of its period.

---

435. What are *variable stars?* How regarded? What difference? What their *period?*
436. What remarkable sample described? Period? Amount of variation? Progress of variation?

437. Another remarkable periodic star is that called *Algol*, in the constellation *Perseus*. It is usually visible as a star of the 2d magnitude, and such it continues for the space of 2 days 14 hours, when it suddenly begins to diminish in splendor; and in about $3\frac{1}{2}$ hours, it is reduced to the 4th magnitude. It then begins again to increase, and in $3\frac{1}{2}$ hours more is restored to its usual brightness; going through all its changes in 2 days 20 hours and 48 minutes, or thereabouts. Through all its successive changes, this star shines with a white light, while the color of all the other variable stars is red.

438. The *cause* of these periodic variations in the brightness of some of the stars is not known. Some suppose them to be occasioned by opake bodies revolving around them, and cutting off a portion of their light from us. Speaking of the sudden obscuration of *Algol*, mentioned above, Dr. Herschel remarks that it indicates a high degree of activity in regions where, but for such evidences, we might conclude all lifeless.

439. "I am disposed," says Dr. Dick, "to consider it as highly probable, that the interposition of the opake bodies of large planets revolving around such stars may, in some cases, account for the phenomena.

"It is true that the planets connected with the solar system are so small, in comparison of the sun, that their interposition between that orb and a spectator, at an immense distance, would produce no sensible affect. But we have no reason to conclude that in all other systems the planets are formed in the same proportions to their central orbs as ours; but from the variety we perceive in every part of nature, both in heaven and earth, we have reason to conclude that every system of the universe is, in some respects, different from another. There is no improbability in admitting that the planets which revolve round some of the stars may be so large as to bear a considerable proportion (perhaps one-half or one-third) to the diameters of the orbs around which they revolve; in which case, if the plane of their orbit lay nearly in a line of our own vision, they would, in certain parts of their revolutions, interpose between our eye and the stars, so as to hide for a time a portion of their surfaces from our view, while in that part of their orbits which is next to the earth."

440. Others, again, are of opinion that those distant suns have one luminous and one opake or clouded hemisphere; and that their variations may thus result from a revolution upon their axes, by which they would present us alternately with their full and their diminished luster.

---

437. What other specimen? Usual appearance? Period? Peculiar color?
438. *Cause* of these variations? Supposition? Dr. Herschel?
439. Dr. Dick's opinion? (Reasoning in note?)
440. What other hypothesis stated?

Another theory is, that these stars are moving with inconceivable velocity in an immense elliptical orbit, the longer axis of which is nearly in a direction toward the eye, and the shorter axis of which would be imperceptible from our system. In such case, the star would appear alternately to approach and recede; now looking in upon our quarter of the universe for a few days, and then rushing back into immensity, to be seen no more by human eyes during the lapse of years or of ages.

441. "Whatever may be the *cause*," says Mr. Abbott, "the fact of these variations is perfectly established, and the contemplation of the stupendous changes which must be occurring in those distant orbs overwhelms the mind with amazement. Worlds vastly larger than our sun suddenly appear, and as suddenly disappear. Now they blaze forth with most resplendent brilliancy, and again they fade away, and often are apparently blotted from existence. These worlds are unquestionably thronged with myriads of inhabitants; and the phenomenon which to us appears but as the waxing or waning luster of a twinkling star, may, to the dwellers on these orbs, be evolutions of grandeur, such as no earthly imagination has ever conceived. But these scenes, now veiled from human eyes, will doubtless all be revealed, when the Christian shall ascend on an angel's wing to the angel's home."

TEMPORARY STARS.

442. Temporary stars are those which have appeared from time to time in different parts of the heavens, blazing forth with extraordinary luster, and, after remaining for a while apparently immovable, died away, and left no traces of their existence behind. Some writers class them among the periodical stars, while others notice them under the head of "New and Lost Stars."

A star of this kind, which appeared in the year 125

---

441. Remarks of Mr. Abbott?
442. What are *temporary stars?* How classed? First noticed? What other instance?

B. C., led Hipparchus to draw up a catalogue of the stars the earliest on record. In A. D. 389, a similar star appeared near the largest star in the Eagle, which, after remaining for three weeks as bright as Venus, disappeared entirely from view.

443. On the 11th of November, 1572, Tycho Brahe, a celebrated Danish astronomer, was returning, in the evening, from his laboratory to his dwelling-house, when he was surprised to find a group of country people gazing upon a star which he was sure did not exist half an hour before. It was then as bright as Sirius, and continued to increase till it surpassed Jupiter in brightness, and was visible at noonday. In December of the same year it began to diminish; and in March, 1574, had entirely disappeared.

This remarkable star was in the constellation *Cassiopeia*, about 5° northeast of the star *Caph*. The place where it once shone is now a dark void!

444. This star was observed for about 16 months, and during the time of its visibility its color exhibited all the different shades of a prodigious flame. "First it was of a dazzling white, then of a reddish yellow, and lastly of an ashy paleness, in which its light expired." "It is impossible," says Mrs. Sumerville, "to imagine any thing more tremendous than a conflagration that could be visible at such a distance."

445. In reference to the same phenomenon, Dr. DICK observes, that "the splendor concentrated in that point of the heavens where the star appeared must have been, in reality, more than equal to the blaze of twelve hundred thousand worlds such as ours, were they all collected into one mass, and all at once wrapped in flames. Nay, it is not improbable, that were a globe as large as would fill the whole circumference of the earth's annual orbit to be lighted up with a splendor similar to that of the

---

443. What other remarkable instance described? By whom? When? In what constellation? Position?
444. How long observed? Appearance? Mrs. Sumerville?
445. Dr. Dick's remarks?

sun, it would scarcely surpass in brilliancy and splendor the star to which we refer."

446. Rev. Prof. VINCE, who has been characterized as "one of the most learned and pious astronomers of the age," advances the opinion, that "the disappearance of some stars may be the destruction of that system at the time appointed by the Deity for the probation of its inhabitants; and the appearance of new stars may be the formation of new systems for new races of beings then called into existence to adorn the works of their Creator."

447. LA PLACE, whose opinion upon such subjects is always entitled to consideration, says: "As to these stars which suddenly shine forth with a very vivid light, and then immediately disappear, it is extremely probable that great conflagrations, produced by extraordinary causes, take place on their surface. This conjecture is confirmed by their change of color, which is analogous to that presented to us on the earth by those bodies which are set on fire, and then gradually extinguished."

448. Dr. JOHN MASON GOODE, author of the *Book of Nature*, &c., seems to have entertained opinions similar to those already expressed. "Worlds and systems of worlds," says he, "are not only perpetually creating, but also perpetually disappearing. It is an extraordinary fact, that within the period of the last century, not less than *thirteen stars*, in different constellations, seem to have totally perished, and *ten new ones* to have been created. In many instances, it is unquestionable, that the stars themselves, the supposed habitations of other kinds or orders of intelligent beings, together with the different planets by which it is probable they were surrounded, have utterly vanished, and the spots they occupied in the heavens have become blanks. What has befallen other systems will assuredly befall our own. Of the *time* and *manner* we know nothing, but the fact is incontrovertible; it is foretold by revelation; it is inscribed in the heavens; it is felt through the earth. Such is the awful and daily text. What, then, ought to be the comment?"

---

446. Prof. Vince's remarks? 447. La Place's? 448. Dr. Goode's?

CHAPTER IV.

CLUSTERS OF STARS AND NEBULÆ.

449. In surveying the concave of the heavens in a clear night, we observe here and there groups of stars, forming bright patches, as if drawn together by some cause other than casual distribution. Such are the *Pleiades* and *Hyades* in Taurus. These are called *Clusters of Stars*. The luminous spot called the *Bee Hive*, in Cancer (visible to the naked eye), is somewhat similar, but less definite, and requires a moderate telescope to resolve it into stars. In the sword-handle of Perseus is another such spot or cluster, which is also visible to the naked eye, but which requires a rather better telescope to resolve it into distinct stars. When fairly in view, however, it is one of the most splendid and magnificent spectacles upon which the eye can rest.

TELESCOPIC VIEW OF THE PLEIADES.

> " O what a confluence of ethereal fires,
> From worlds unnumber'd down the steep of heaven,
> Stream to a point, and center on my sight."

450. Many of these faint and compact clusters have been mistaken for comets, as through telescopes of mod-

---

449. Clusters? Specimens?
450. What mistake respecting? What like? How known that they are not comets?

erate power they appear like such. Messier has given a list of 103 objects of this sort, with which all who search for comets ought to be familiar, to avoid being misled by their similarity of appearance. That they are not comets, is evident from their fixedness in the heavens, and from the fact, that when we come to examine them with instruments of great power, they are perceived to consist entirely of stars, crowded together so as to exhibit a definite outline, and to run up to a blaze of light in the center, where their condensation is usually the greatest.

451. Some of these clusters are of an exceedingly rough figure, and convey the idea of a globular space filled full of stars, insulated in the heavens, and constituting in itself a family or society apart from the rest, and subject only to its own internal laws.

ROUND CLUSTER IN CAPRICORN.

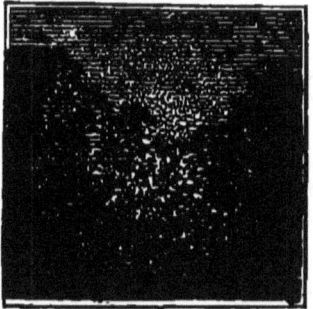

It would be a vain effort to attempt to count the stars in one of these clusters. They are not to be reckoned by hundreds; and on a rough calculation, grounded on the apparent intervals between them at the borders, and the angular diameter of the whole group, it would appear that many clusters of this description must contain, at least, from ten to twenty thousand stars, compacted and wedged together in a round space, whose angular diameter does not exceed eight or ten minutes, or an area equal to a tenth part of that covered by the moon.

452. Some of these clusters have a very irregular outline. These are generally less rich in stars, and especially less condensed toward the center. They are also less definite in point of outline. In some of them, the stars are nearly all of a size; in others, extremely different. It is no uncommon thing to find a very red star, much

---

451. What said of the *form* of these clusters? Stars in each? Apparent diameter?

452. What further respecting forms? Character of irregular clusters?

brighter than the rest, occupying a conspicuous situation in them.

453. It is by no means improbable that the individual stars of these clusters are suns like our own, the centers of so many distinct systems, and that their mutual distances are equal to those which separate our sun from the nearest fixed stars. Besides, the round figure of some of these groups seems to indicate the existence of some general bond of union, of the nature of an attractive force.

RICH CLUSTER IN BERENICES' HAIR.

This is one of the most gorgeous clusters in all the heavens. Sir John Herschel pronounced it the most magnificent object he had ever beheld. It is about 6' in diameter, and contains a countless throng of stars, that scarcely ever fail to elicit a burst of surprise and astonishment from the beholder! Who can gaze upon such a scene, and not for a time forget earth, in the rapt contemplation of the distant glory?

"There's not a scene to mortals given,
That more divides the soul and clod,
Than yon proud heraldry of heaven—
You burning blazonry of God."

A similar cluster, though somewhat different in form, may be found between ζ and η, in Hercules. This, too, is a most magnificent object. Under favorable circumstances, it may be seen with the naked eye; and by the aid of telescopes, it is easily resolved into myriads of stars. "It is, indeed, truly glorious," says Smyth, "and enlarges on the eye by studious gazing." "Perhaps," says Prof. Nichol, "no one ever saw it, for the first time, through a telescope, without uttering a shout of wonder."

NEBULÆ.

454. The term *Nebulæ* is applied to those clusters of stars that are so distant as to appear only like a faint cloud or haze of light. In this sense, some of the clusters heretofore described may be classed as nebulæ; and, indeed, it may be said of all the different kinds of nebulæ, that it is impossible to say where one species ends, and another begins.

---

453. What said of individual stars in clusters? Of round figure of some clusters? (What specimen in cut? What said of it? Angular diameter? Effect of seeing? Poetry? What other similar cluster? What said of it?)
454. What are *Nebulæ*? How differ from clusters?

455. *Resolvable Nebulæ* are those clusters, the light o whose individual stars are blended together, when seer. through a common telescope; but which, when viewed through glasses of sufficient power, can be resolved into distinct stars.

456. *Irresolvable Nebulæ* are those nebulous spots which were formerly supposed to consist of vast fields of matter in a high state of rarefaction, and not of distinct stars. But it is doubtful whether any nebulæ exist which could not be resolved into stars, had we telescopes of sufficient power.

"About the close of last year," says Dr. Scoresby, in 1846, "the Earl of Rosse succeeded in getting his great telescope into complete operation; and during the first month of his observations on fifty of the unresolvable nebulæ, he succeeded in ascertaining that 43 of them were already resolvable into masses of stars. Thus is confirmed the opinion, that we have only to increase the power of the instrument to resolve all the nebulæ into stars, and the grand nebular hypothesis of La Place into splendid astronomical dream."

DOUBLE NEBULÆ.

457. Nebulæ of almost every conceivable shape may be found in the heavens. Some are round—others elliptical. Some occur singly, while others are double, or seem to be connected together.

The specimen here shown is in the Greyhound. The two nebulæ are elliptical, as shown, and are so united as to stand perpendicularly to each other.

458. *Annular Nebulæ* are those that exhibit the form of a *ring*. Of these, but few specimens are known. One of the most striking may be found about 6° below *Mizar*,

---

455. What are resolvable nebulæ? How when seen through powerful telescopes?
456. Irresolvable nebulæ? Are any nebulæ really irresolvable? Remarks from Dr. Scoresby?
457. What further description of nebulæ? Specimen?
458. What are *annular* nebulæ? Are they common? What specimen is cut? Describe it.

the middle star in the tail of the Great Bear. It consists of a large and bright globular nebula, surrounded by a double ring, at a considerable distance from the globe; or rather a single ring divided through about two-fifths of its circumference, and having one portion turned up, as it were, out of the plane of the rest.* A faint nebulous atmosphere, and a small round nebula near it, like a satellite, completes the figure.

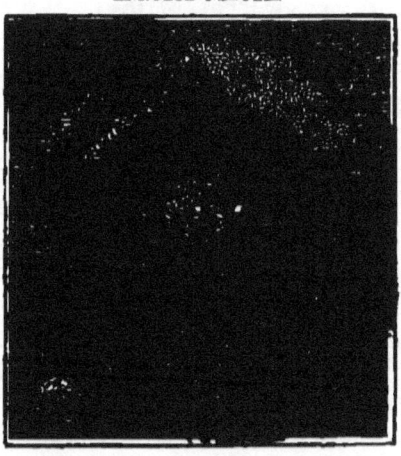

ANNULAR NEBULÆ.

459. Another very conspicuous nebula of this class may be found half-way between β and γ, in the Lyre, and may be seen with a telescope of moderate power. It is small, and particularly well defined, so as, in fact, to have much more the appearance of a flat oval solid ring, than of a nebula. The space within the ring is filled with a faint hazy light, uniformly spread over it, like a fine gauze stretched over a hoop.

460. "*Planetary Nebulæ*," says Dr. Herschel, "are very extraordinary objects. They have, as their name imports, exactly the appearance of planets—round or slightly oval discs—in some instances quite sharply terminated, in others a little hazy at the borders, and of a light exactly equable, or only a very little mottled, which, in some of them, approaches in vividness to that of the actual planets. Whatever be their nature, they must be of enormous magnitude."

461. *Stellar Nebulæ*, or *Nebulous Stars*, are such as present the appearance of a thin cloud, with a bright star in or near the center. They are round or oval-

---

459. What other annular nebulæ? Describe.
460. *Planetary* nebulæ? Describe.
461. *Stellar* nebulæ? Remarks of Professor Mitchel?

shaped, and look like a star with a burr around it, or a candle shining through horn. "It was an object of this kind," says Prof. Mitchel, "which first suggested to Sir W. Herschel his great theory of the formation of suns out of a nebulous fluid. He thought it impossible to account for the central location of stars, surrounded by nebulous matter, in any way except by supposing this to be a sort of atmosphere attracted to, and sustained in its spherical form by, the power of the central body. I have examined specimens of these objects, and always with increasing wonder. Their magnitude must be enormous, as the stars are certainly not nearer than other stars; and yet the circular halo around them is of a diameter easily measured, and proves them to have a circumference perhaps greater than the entire orbit of Neptune."

STELLAR NEBULÆ.

462. One of the most remarkable nebula in all the heavens may be found around the middle star in the sword of Orion. It is easily seen with a common telescope. It is shaped like the head of some animal—a fish, for instance—with its mouth open. Near the inner surface of this mouth are four stars, ranged in the form of a *trapezium*. It requires a good telescope to see four stars; but, with powerful instruments, six are visible, instead of four.

GREAT NEBULA IN ORION.

---

462. Describe the nebula of Orion? Where situated? Shape? What stars in it?

463. The *sun* is considered by astronomers as belonging to this class of nebulous stars; and the *Zodiacal Light* (322 and 325) has been regarded as of the nature of the gaseous matter with which the nebulous stars are surrounded. It is supposed that if we were as far from the sun as from the stellar nebulæ, he would appear to us only as a small and nebulous star!

464. Until recently, the most powerful instruments have failed to reveal any thing like distinct stars, as composing the body of the remarkable nebula in Orion. Both the Herschels regarded it as positively irresolvable; or, in other words, as composed of nebulous fluid or unorganized matter. But it has recently been seen to be composed of distinct stars, both by the monster telescope of Lord Rosse, and the great refractor of Cambridge, near Boston.

465. The *magnitude* of this nebula must be beyond all human conception. "If," says Mr. Smyth, "the parallax of this nebula be no greater than that of the stars, its breadth cannot be less than a hundred times that of the diameter of the earth's orbit; but if, as is more probable, it is a vast distance beyond them, its magnitude must be utterly inconceivable."

466. *Prof. Mitchel* observes, that in case light be not absorbed in its journey through the celestial spaces, the light of the nebula of Orion cannot reach the eye in less than 60,000 years, with a velocity of twelve millions of miles in every minute of time! And yet this object may be seen from this stupendous distance, even by the naked eye! What, then, must be its dimensions? Here, indeed, we behold a *universe* of itself too vast for the imagination to grasp, and yet so remote as to appear a faint spot upon the sky."

467. The *number* of such nebulous bodies is unknown

---

463. Remarks respecting the sun?
464. How the nebula in Orion regarded? What recent discovery?
465. Its probable magnitude? Remark of Smyth?
466. Prof. Mitchel's observations respecting its distance and dimensions?
467. What said of the *number* of nebulous bodies in the heavens? Where most abundant? Herschel's catalogue? Various forms?

perhaps we should say innumerable. They are especially abundant in the *Galaxy* or Milky Way. Sir W. Herschel arranged a catalogue, showing the places of *two thousand* of these objects. They are of all shapes and sizes, and of all degrees of brightness, from the faintest milky appearance to the light of a fixed star.

468. *Star Dust* is a name given to those exceedingly faint nebulous patches that appear to be scattered about at random in the far-distant heavens. It is barely visible through the best telescopes, and seems to form a sort of back-ground, far beyond all stars, clusters, and nebulæ, resolvable or irresolvable.

469. "The nebulæ," says Sir John Herschel, "furnish, in every point of view, an inexhaustible field of speculation and conjecture. That by far a larger share of them consist of stars, there can be little doubt; and in the interminable range of system upon system, and firmament upon firmament, which we thus catch a glimpse of, the imagination is bewildered and lost."

470. It is a general belief among astronomers that the material universe consists of distinct clusters, separated from each other by innumerable chasms: that the fixed stars by which we are surrounded constitute *one great cluster*—the sun being a star with the rest, and appearing as he does to us, solely on account of our nearness to him; that the nebulæ are far beyond our cluster, like so many distinct continents in the boundless ocean of immensity.

471. Could we leave our system, and pass outward toward the fixed stars, they would doubtless expand to the dimensions of suns as we approached them, while our own central luminary would dwindle to a glimmering star. Reaching the frontier of the cluster, and plunging off into the awful solitudes of space, toward the distant nebulæ beyond, we should see them also expand as we drew near, while our vast firmament of stars seemed to

---

468. What is meant by *star dust?* Where supposed to be situated?
469. Herschel's remark respecting the nebulæ?
470. What the prevailing opinion among astronomers, as to the structure of the universe?
471. What imaginary journey and scenery described by the author?

be gathering into a compact group; till at length, entering the bosom of the distant nebulæ, we should find ourselves surrounded by new and strange constellations; and if we saw our own firmament at all, should see it only as a faint annular nebula, far beyond the reach of all unassisted vision.

472. The great stellar cluster in which the sun and solar system are imbedded is supposed, in its form, to resemble a double convex lens, with the sun and solar system near its center; and by being viewed *edgewise* from our central position, to give us the phenomenon of the *Milky Way*.

GREAT NEBULA OF THE SOLAR SYSTEM.

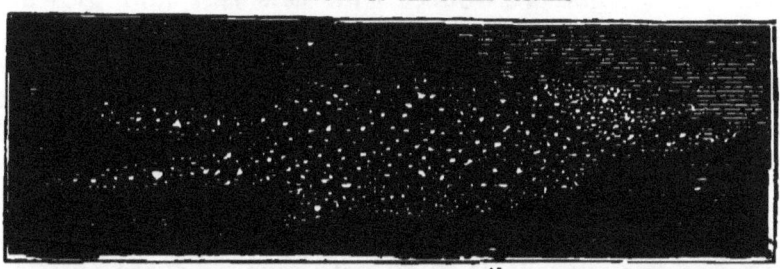

The above is an edgewise view of the great stellar cluster, in the midst of which the solar system is placed, as drawn by Sir William Herschel. Its figure was ascertained by gauging the space-penetrating power of his telescope, and then "sounding the heavens," to ascertain the distance *through* the cluster, in all directions, to the open void. The nebulæ lie in distinct and independent islands, far beyond the limits of our cluster.

Let the student imagine the sun to be one of the stars near the middle of the lens-shaped cluster, of which the above is an edge view, with the planets revolving close around it. If, then, he look out upon the surrounding stars, the number visible, and their distinctness, will depend upon the direction in which he looks. If toward the thin part of the cluster (either up or down in the cut), fewer stars will be seen, while they will be comparatively distinct. But if the view be toward the *edge* of the cluster, instead of the sides (or horizontally, in the cut), there will be seen beyond the large stars, and fading away to an indistinct and mingled light, a numberless host of stars; and this zone of distant stars will extend quite around the heavens. Such is the Galaxy or Milky Way. The zone of milky light is the light of the stars in the remote edge of the great cluster. The opening in the left end of the figure is a split in the cluster, and constitutes the division seen in the milky way, extending part way around the heavens. See cut page 203.

The vast apparent *extent* of the Galaxy, as compared with other nebulæ, is supposed to be justly attributable to its comparative *nearness*. Were we as far from the solar system as from the nebulæ in the Lyre, the Milky Way would doubtless appear as an *annular nebula* no larger than that. It may therefore with propriety be called "the great nebula of the solar system."

473. Sir W. Herschel estimated that 50,000 stars passed the field of his telescope, in the Milky Way, in a

---

472. Supposed form of our own stellar cluster? Philosophy of Galaxy? (Why apparently so large? How appear at a great distance?)
473. Stars in Milky Way? Mutual distances? Character of each star?

single hour! And yet the space thus examined was hardly a point in the mighty concave of our own "sun-strown firmament." What an idea is here conveyed to the mind, of the almost boundless extent of the universe! The mutual distances of these innumerable orbs are probably not less than the distance from our sun to the nearest fixed stars, while they are each the center of a distinct system of worlds, to which they dispense light and heat.

474. Were the universe limited to the Great Solar Cluster, in the midst of which we are placed, it would be impossible to conceive of its almost infinite dimensions; but when we reflect that this vast and glowing zone of suns is but one of thousands of such assemblages, which, from their remoteness, appear only as fleecy clouds hovering over the frontiers of space, we are absolutely overwhelmed and lost in the mighty abyss of being!

475. And here we close our rapid and necessarily imperfect survey of the Sidereal Heavens. And while the mind of the student is filled with awe, in contemplating the vastness and majesty of creation, let him not forget that over all these Jehovah reigns—that "these are but parts of his ways;" and yet so perfect is his knowledge and providence in every world, that the very hairs of our heads are numbered, and not a sparrow falls without his notice. And while we behold the wisdom, power, and goodness of God so gloriously inscribed in the heavens, let us learn to be humble and obedient—to love and serve our Maker here—that we may be prepared for the still more extended scenes of another life, and for the society of the wise and good in a world to come.

---

474. Magnitude of our own cluster? What in comparison with all others?
475. Remarks in closing paragraph? Moral reflections?

PART III.

PRACTICAL ASTRONOMY.

CHAPTER I.

PROPERTIES OF LIGHT.

476. *Practical Astronomy* has respect to the *means* employed for the acquisition of astronomical knowledge. It includes the properties of light, the structure and use of instruments, and the processes of mathematical calculation.

In the present treatise, nothing further will be attempted than a mere introduction to practical astronomy. In a work designed for popular use, mathematical demonstrations would be out of place. Still, every student in astronomy should know how telescopes are made, upon what laws they depend for their power, and how they are used. It is for this purpose mainly that we add the following chapters on Practical Astronomy.

477. *Light* is that invisible ethereal substance by which we are apprised of the existence, forms, and colors of material objects, through the medium of the visual organs. To this subtile fluid we are especially indebted for our knowledge of those distant worlds that are the principal subjects of astronomical inquiry.

478. The term *light* is used in two different senses. It may signify either light itself, or the *degree* of light by which we are enabled to see objects distinctly. In this last sense, we put light in opposition to darkness. But

---

476. Parts of the book gone over? Subject of Part III.? Of Chapter I.? What is *practical astronomy?* (How far discussed in this treatise?)
477. Define light. For what indebted to it?
478. Different senses in which the term is used? What is darkness? Can it be dark and light at the same time? Is there any place without light? (Quotation from Dick?)

it should be borne in mind that *darkness* is merely the absence of that *degree* of light which is necessary to human vision; and when it is dark to us, it may be light to many of the lower animals. Indeed, there is more or less light even in the darkest night, and in the deepest dungeon.

"Those unfortunate individuals," says Dr. Dick, "who have been confined in the darkest dungeons, have declared, that though, on their first entrance, no object could be perceived, perhaps for a day or two, yet, in the course of time, as the pupils of their eyes expanded, they could readily perceive mice, rats, and other animals that infested their cells, and likewise the walls of their apartments; which shows that, even in such situations, light is present, and produces a certain degree of influence."

479. Of the *nature* of the substance we call light two theories have been advanced. The first is, that the whole sphere of the universe is filled with a subtile fluid, which receives from luminous bodies an *agitation*; so that, by its continued vibratory motion, we are enabled to perceive luminous bodies. This was the opinion of Descartes, Euler, Huygens, and Franklin.

The second theory is, that light consists of particles thrown off from luminous bodies, and actually proceeding through space. This is the doctrine of Newton, and of the British philosophers generally.

Without attempting to decide, in this place, upon the relative merits of these two hypotheses, we shall use those terms, for convenience sake, that indicate the actual passage of light from one body to another.

480. Light proceeds from luminous bodies in straight lines, and in all directions. It will not wind its way through a crooked passage, like sound; neither is it confined to a part of the circumference around it.

As the sun may be seen from every point in the solar system, and far hence into space in every direction, even till he appears but a faint and glimmering star, it is evident that he fills every part of this vast space with his beams. And the same might be said of every star in the firmament.

481. As vision depends not upon the *existence* of light merely, but requires a certain *degree* of light to emanate from the object, and to enter the pupil of the eye, it is obvious that if we can, by any means, *concentrate* the

---

479. What theories of the *nature* of light, and by whom supported respectively? (Remark of author?)
480. How light proceeds from luminous bodies? (Radiations from sun and stars?)
481. How improve vision, and why? (Animals?)

light, so that more may enter the eye, it will improve our perception of visible objects, and even enable us to see objects otherwise wholly invisible.

Some animals have the power of adapting their eyes to the existing degree of light. The cat, horse, &c., can see day or night; while the owl, that sees well in the night, sees poorly in the day-time.

482. Light may be turned out of its course either by *reflection* or *refraction*. It is *reflected* when it falls upon the highly polished surface of metals and other intransparent substances; and *refracted* when it passes through transparent substances of different densities.

REFRACTION OF LIGHT.

483. Whenever light passes from a rare medium to one more dense, and enters the latter obliquely, it invariably leaves its first direction, and assumes a new one. This change or bending of the rays of light is what is called *Refraction*.

The term *refract* is from the Latin *re*, and *frango*, to break; and signifies the breaking of the natural course of the rays.

484. As *air* and *water* are both transparent, but of different densities, it follows that, when light passes obliquely from one to the other, it will be refracted. If it pass from the air into the water, it will be refracted *toward a perpendicular*.

LIGHT REFRACTED BY WATER.

Here the ray A C strikes the water perpendicularly, and passes directly through to B without being refracted. But the ray D C strikes the water at C obliquely; and instead of passing straight through to E, is refracted at C, and reaches the bottom of the water at F. If, therefore, a person were to receive the ray into the eye at F, and to judge of the place of the object from which the light emanates from the direction of the ray C F, he would conclude that he saw the object at G, unless he made allowance for the refraction of the light at C.

482. How light turned out of course?
483. What is *refraction?* How produced? (Derivation of term *refract?*)
484. How refracted by air and water? (Illustrate by diagram.)

485. When light passes obliquely from a denser to a rarer medium, as from water into air, it is refracted *from* a perpendicular toward a horizontal.

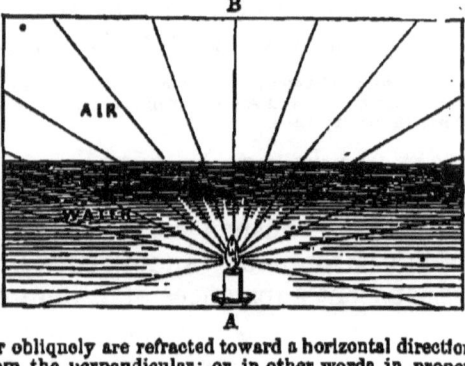

LIGHT PROCEEDING FROM WATER.

Here the lamp A shines up through water into air. The ray that strikes the surface perpendicularly passes on to B without being refracted; but the other rays that leave the water obliquely are refracted toward a horizontal direction, in proportion to their distance from the perpendicular; or, in other words, in proportion to the obliquity of their contact with the surface of the water.

486. In consequence of the refraction of light toward a horizontal direction, in passing from water into air, a pole, half of which is in the water, seems bent at the surface, and the lower end seems nearer the surface than it really is. For the same reason, the bottom of a river seems higher, if seen obliquely, than it really is; and the water is always deeper than we judge it to be.

EFFECT OF REFRACTION.

In this cut, the *oar*, the blade of which is in the water, seems bent at the surface of the water. The rays of light passing from the part under water to the surface at D, are refracted toward a horizontal direction at that point, and received into the eye of the observer at B, who, judging of the position of the immersed portion of the oar from the direction of the rays D B, locates the blade of the oar at C; thus reversing the effect illustrated at 484.

487. The refracting power of different transparent substances depends mainly upon their density. Water refracts more than air, glass more than water, and diamond most of all. But the angle of incidence, or the obliquity of the contact of the rays with the denser sub-

---

485. How when light passes from denser to rarer mediums? (Diagram.)
486. Effect of refraction upon objects seen under water? (Diagram.)
487. Upon what does the refracting power of different transparent media epend?

stance, has also much to do in determining the amount of refraction.

488. By the aid of refraction, we may see objects that are actually *behind* an opake or intransparent body.

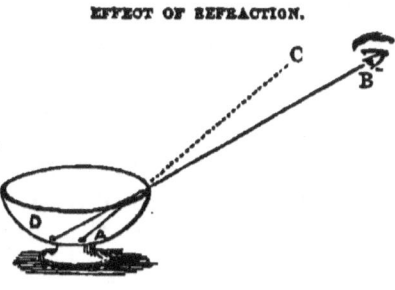

EFFECT OF REFRACTION.

Here the piece of money at A, at the bottom of the cup, would be invisible to the beholder at B, if the cup was empty, as the light from the money would pass from A to C; but when the cup is filled with water, the light is refracted to B, and the beholder sees the money apparently at D.

489. By the law of refraction, light has been found to consist of a combination of colors. By passing a beam of light through a triangular piece of flint glass called a *prism*, it is seen that some parts of the light are more refrangible than others, so that the light is analyzed, or separated into its component parts or elements.

REFRACTION BY A PRISM.

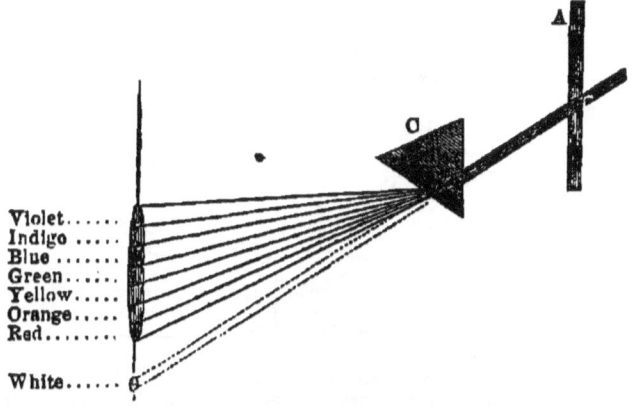

Let a ray of light from the sun be admitted through a hole in the window shutter, A into a room from which all other light is excluded; it will form, on a screen placed a little distance in front, a circular image, B, of white light. Now, interpose near the shutter a glass prism, C, and the light, in passing through it, will not only be refracted in the same direction, both when it enters the prism and when it leaves it, but the several rays of which white light is composed will be separated, and will arrange in regular order on the screen, immediately above the image B, which will disappear. The violet ray, it will be seen, is most refracted, and the red least; the whole forming on the screen an elongated image of the sun, called the *solar spectrum.—Johnston.*

488. What other effect of refraction? (How illustrated?)
489. What discovery by refraction? (How made?)

490. It is the refraction of the clouds that gives the sky its beautiful colors morning and evening; and the refracting power of the rain-drops produces the beautiful phenomenon of the rainbow.

ATMOSPHERICAL REFRACTION.

491. The refracting power of the atmosphere produces many curious phenomena. Sometimes ships are seen bottom upwards in the air, single or double. At other times, objects really below the horizon, as ships or islands, seem to rise up, and to come distinctly in view.

492. A very important effect of refraction, as it relates to astronomy, is, that it more or less affects the apparent places of all the heavenly bodies. As the light coming from them strikes the atmosphere obliquely, and passes downward through it, it is refracted or bent toward the earth, or toward a perpendicular. And as we judge of the position of the object by the direction of the ray when it enters the eye, we place objects higher in the heavens than they really are.

ATMOSPHERICAL REFRACTION.

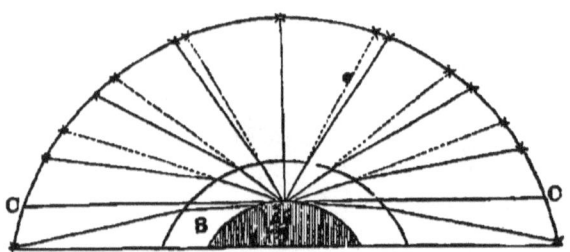

Let A, in the cut, represent the earth; B, the atmosphere; C C, the visible horizon; and the exterior circle the apparent concave of the heavens. Now, as the light passes from the stars, and strikes the atmosphere, it is seen to curve downward, because it strikes the atmosphere obliquely; and the air increases in density as we approach the earth. But as the amount of refraction depends not only upon the density, but also upon the obliquity of the contact, it is seen that the refraction is greatest at the horizon, and gradually diminishes till the object reaches the zenith, when there is no obliquity, and the refraction wholly ceases. The dark lines in the cut show the true, and the dotted the apparent positions.

---

490. What other effects of refraction?
491. Atmospherical refraction? Effects on terrestrial objects?
492. Upon apparent places of stars, &c.? (Diagram. What said of exaggeration?)

In the cut, the depth of the atmosphere, as compared with the globe, is greatly exaggerated. Even allowing it to be 50 miles deep, it is only $\frac{1}{80}$th of the semi-diameter of the globe, which is equal to only about $\frac{1}{12}$th of an inch upon a common 18-inch globe. But it was necessary to exaggerate, in order to illustrate the principle.

493. The amount of displacement of objects in the horizon, by atmospherical refraction, is about 33', or a little more than the greatest apparent diameter of either the sun or moon. It follows, therefore, that when we see the lower edge of either *apparently* resting on the horizon, its whole disk is in reality below it; and would be entirely concealed by the convexity of the earth, were it not for refraction.

494. Refraction sometimes causes the sun and moon to appear elongated horizontally, when near the horizon, and seen through a dense atmosphere. The rays from their lower limb being refracted more than those from the upper limb, on account of coming to us through a lower and denser portion of the atmosphere, the lower portion seems higher in proportion; or, in other words, the perpendicular diameter of the object seems the shortest. It is then called a *horizontal moon*.

495. Another effect of refraction is, that the sun seems to arise about three minutes earlier, and to set about three minutes later, on account of atmospherical refraction, than it otherwise would; thus adding about six minutes, on an average, to the length of each day.

The atmosphere is said to be so dense about the North Pole as to bring the sun above the horizon some days before he should appear, according to calculation. In 1596, some Dutch navigators, who wintered at Nova Zembla, in latitude 76°, found that the sun began to be visible 17 days before it should have appeared by calculation; and Kepler computes that the atmospheric refraction must have amounted to 5°, or 10 times as much as with us.

496. The *twilight* of morning and evening is produced partly by refraction, but mainly by reflection. In the morning, when the sun arrives within 18° of the horizon, his rays pass over our heads into the higher region of the atmosphere, and are thence reflected down to the earth. The day is then said to *dawn*, and the light gradually increases till sunrise. In the evening, this

---

*493. Amount of displacement of celestial objects by refraction? What follows?
494. What effect upon apparent form of moon, &c.?
495. On length of days? (How about North Pole?)
496. Cause of *twilight*? (Note.)

process is reversed, and the twilight lingers till the sun is 18° below the horizon. There is thus more than an hour of twilight both morning and evening.

<small>In the arctic regions, the sun is never more than 18° below the horizon; so that the twilight continues during the whole night.</small>

497. In making astronomical observations, for the purposes of navigation, &c., allowance has to be made for refraction, according to the altitude of the object, and the state of the atmosphere. For this purpose tables are constructed, showing the amount of refraction for every degree of altitude, from the horizon to the zenith.

REFRACTION BY GLASS LENSES.

498. A *lens* is a piece of glass or other transparent substance, of such a form as to collect or disperse the rays of light that are passed through it, by refracting them out of a direct course. They are of different forms, and have different powers.

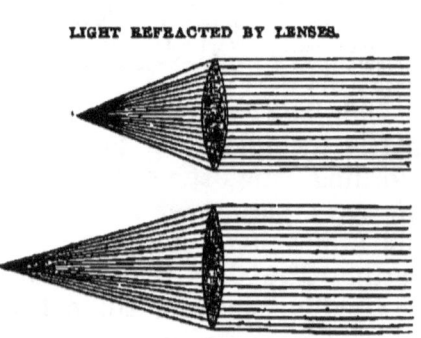

LENSES OF DIFFERENT FORMS.

<small>In the adjoining cut, we have an edgewise view of six different lenses.

A is a *plano-convex*, or half a double convex lens; one side being convex, and the other plane.

B is a *plano-concave*; one surface being concave, and the other plane.

C is a *double-convex* lens, or one that is bounded by two convex surfaces.

D is a *double-concave* lens, or a circular piece of glass hollowed out on both sides.

E is a *concavo-convex* lens, whose curves differ, but do not meet, if produced.

F is a *meniscus*, or a concavo-convex lens, the curves of whose surfaces meet.</small>

499. A *double-convex* lens converges parallel rays to a point called the *focus;* and the distance of the focus depends upon the degree of convexity.

LIGHT REFRACTED BY LENSES.

<small>In the first of these cuts, the lens is quite thick, and the focus of the rays is quite near; but the other being less convex, the focus is more remote.</small>

<small>497. What allowance for refraction? Tables?
498. What is a *lens?* (Draw and describe different kinds?)
499. Refracting power of *double-convex* lens? Focal distance? (Diagram and illustrate.)</small>

500. The distance of the focus of a *double-convex* glass lens is the radius of the sphere of its convexity.

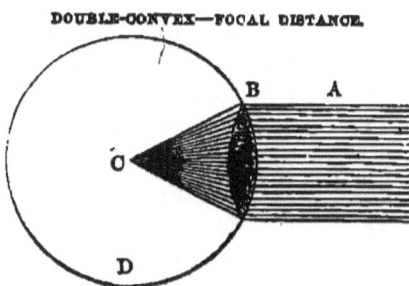
DOUBLE-CONVEX—FOCAL DISTANCE.

In this cut, it will be seen that the parallel rays A are refracted to a focus at C, by the double convex lens B, the convexity of whose surfaces is just equal to the curve of the circle D.

501. The focal distance of a *plano-convex* lens is equal to the diameter of the sphere formed by the convex surface produced.

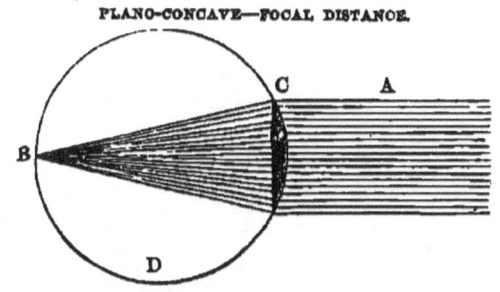
PLANO-CONCAVE—FOCAL DISTANCE.

It must be borne in mind that light is refracted both when it enters and when it leaves a double-convex lens, and in both instances in the same direction; and, so far as the distance of the focus is concerned, to the same extent. But when the lens is convex only on one side, half its refracting power is gone, so that the rays are not so soon refracted to a focus. In this case, the focal distance is equal to the *diameter* of the sphere formed by extending the convex surface of the lens; while with the double-convex lens, the focal distance is only equal to the *radius* of such sphere. In the cut, the parallel rays A are refracted to a focus at B, by the plano-concave lens C; and the distance C B is the diameter of the circle D, formed by the convex surface of the lens C produced.

502. A *double-concave* lens *disperses* parallel rays, as if they diverged from the center of a circle formed by the convex surface produced.

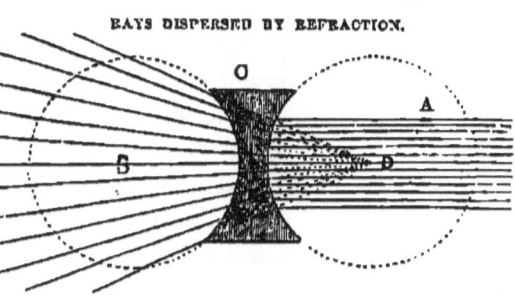
RAYS DISPERSED BY REFRACTION.

In this cut, the parallel rays A are dispersed by the double-concave lens B, as shown at C; and their direction, as thus refracted, is the same as if they proceeded from the point D, which is the center of a circle formed by the concave surface of the lens produced.

500. How focal distance governed? (Diagram.)
501. What is the focal distance of a *plano-convex* lens? (Diagram.)
502. Effect of *double-convex* lens? Amount of divergency of rays? (Diagram.)

503. Common spectacles, opera-glasses, burning-glasses, and refracting telescopes are made by converging light to a focus, by the use of double-convex lenses.

The ordinary burning-glass, which may be bought for a few shillings, is a double-convex disk of glass two or three inches in diameter, inclosed in a slight metallic frame, with a handle on one side. Old tobacco-smokers sometimes carry them in their pockets, to light their pipes with when the sun shines. In other instances, they have been so placed as to fire a cannon in clear weather, by igniting the priming at 12 o'clock.

BURNING-GLASS.

The adjoining cut represents a large burning-glass converging the rays of the sun to a focus, and setting combustible substances on fire. Such glasses have been made powerful enough to melt the most refractory substances, as platinum, agate, &c. "A lens three feet in diameter," says Professor Gray, "has been known to melt carnelian in 75 seconds, and a piece of white agate in 30 seconds."

REFLECTION OF LIGHT.

504. We have now shown how light may be turned out of its course, and analyzed, dispersed, or converged to a point by *refraction*. Let us now consider how it may be converged to a focus by *reflection*.

505. When light falls upon a highly polished surface, especially of metals, it is *reflected* or thrown off in a new direction, and the angles of contact and departure are always equal.

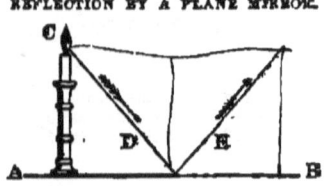

REFLECTION BY A PLANE MIRROR.

Let A B represent the polished metallic surface. C the source of light, and the arrows the direction of the ray. Then D would represent the angle of incidence or contact, and E the angle of reflection or departure—which angles are seen to be equal.

---

503. What articles made with double-convex lenses? Uses? (Power of burning-glasses?)
504. What now shown in this chapter? What next?
505. What is *reflection*, and when does it take place? What law governs it? (Diagram.)

506. A *concave mirror* reflects parallel rays back to a focus, the distance of which is equal to half the radius of the sphere formed by the concave surface produced.

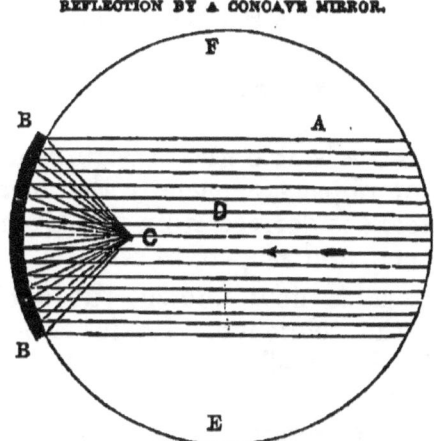

REFLECTION BY A CONCAVE MIRROR.

In this cut, the parallel rays A fall upon the concave mirror B B, and are reflected to the focus C, which is half the radius of the sphere formed by the surface of the mirror produced. If, therefore, it was desirable to construct a concave mirror, having its focus 10 feet distant, it would only be necessary to grind it on the circle of a sphere having a radius of 20 feet.

507. In reflection, a portion of the light is absorbed or otherwise lost, so that a reflector of a given diameter will not converge as much light to a focus as a double-convex lens of the same size. In the latter case, all the light is transmitted. Still, reflectors have been formed of such power as to melt iron, and other more difficult substances.

We have now considered so much of optics as is necessary to an understanding of the principles upon which telescopes are constructed; and, for further particulars, shall refer the student to books of Natural Philosophy.

---

506. How does a *concave* mirror reflect parallel rays? Distance of focus? (Diagram. How would you construct a concave mirror with a 10 feet focus?)
507. Is all the light falling upon a polished surface reflected? What then? (Closing note?)

CHAPTER II.

TELESCOPES.

508. A *Telescope* is an optical instrument employed in viewing distant objects, especially the heavenly bodies. The term *telescope* is derived from two Greek words, viz., *tele*, at a distance, and *skopeo*, to see.

509. So far as is now known, the ancients had no knowledge of the telescope. Its invention, which occurred in 1609, is usually attributed to *Galileo*, a philosopher of Florence, in Italy.

<small>The discovery of the principle upon which the refracting telescope is constructed was purely accidental. The children of one *Jansen*, a spectacle-maker of Middleburgh, in Holland, being at play in their father's shop, happened to place two glasses in such a manner, that in looking through them, at the weather-cock of the church, it appeared to be nearer and much larger than usual. This led their father to fix the glasses upon a board, that they might be ready for observation; and the news of the discovery was soon conveyed to the learned throughout Europe. Galileo hearing of the phenomenon, soon discovered the secret, and put the glasses in a *tube*, instead of on a board; and thus the first telescope was constructed.</small>

510. The telescope of Galileo was but one inch in diameter, and magnified objects but 30 times. Yet with this simple instrument he discovered the face of the moon to be full of inequalities, like mountains and valleys; the spots on the sun; the phases of Venus; the satellites of Jupiter; and thousands of new stars in all parts of the heavens.

<small>Notwithstanding this propitious commencement, so slow was the progress of the telescope toward its present state, that in 1816, Bonnycastle speaks of the 30-fold magnifying power of the telescope of Galileo as "nearly the greatest perfection that this kind of telescope is capable of!"</small>

511. If he be the real author of an invention who, from a knowledge of the cause upon which it depends, deduces it from one principle to another, till he arrives

---

<small>508. Subject of Chap. II.? Telescope? Derivation?
509. Ancient or modern? Inventor? (Incidents of discovery?)
510. Galileo's telescope? Discoveries with it? (Progress in telescope making?)
511. Is Galileo entitled to the honor of inventing the telescope? (Where is his?)</small>

at the end proposed, then the whole merit of the invention of the telescope belongs to *Galileo*. The telescope of *Jansen* was a rude instrument of mere curiosity, accidentally arranged; but *Galileo* was the first who constructed it upon principles of science, and showed the practical uses to which it might be applied.

It is said that the original telescope constructed by Galileo is still preserved in the British Museum. A pigmy, indeed, in its way, but the honored progenitor of a race of giants!

512. The discovery of the telescope tended greatly to sustain the Copernican theory, which had just been promulgated (10), and of which Galileo was an ardent disciple. Like Copernicus, however, his doctrines subjected him to severe persecutions, and he was obliged to renounce them.

The following is his renunciation, made June 28, 1633: "I, Galileo, in the seventieth year of my age, on bended knees before your eminences, having before my eyes and touching with my hands the Holy Gospels, I curse and detest the error of the earth's movement." As he left the court, however, after this forced renunciation, he is said to have stamped upon the earth, and exclaimed, "It does move, after all!" Ten years after this he was sent to prison for the same supposed error; and soon, his age advancing, the grave received him from the malice of his persecutors.

DIFFERENT KINDS OF TELESCOPES.

513. Telescopes are of two kinds—*Reflectors* and *Refractors*. Refracting telescopes are made by *refracting* the light to a focus with a glass lens (499); and reflecting telescopes, by *reflecting* it to a focus with a concave mirror (506). Besides this general division, there are various kinds, both of reflectors and refractors.

514. Telescopes assist vision in various ways—first, by enlarging the visual angle under which a distant object is seen, and thus magnifying that object; and, secondly, by converging to a point more light than could otherwise enter the eye—thus rendering objects distinct or visible that would otherwise be indistinct or invisible.

All the light falling upon a six or a twelve inch lens may be converged to a focus, so as to be taken into the human eye through the pupil, which is but one-fourth of an inch in diameter. Our vision is thus made as perfect by art as if nature had given us ability to enlarge the eye till the pupil was a foot in diameter.

---

512. Relation of discovery to Copernican theory? Effects upon Galileo? (His renunciation? Death?)
513. Kinds of telescopes? Describe.
514. How assist vision? (Illustrative note?)

515. Refracting telescopes may consist of a double-convex lens placed upon a stand, without tube or eye-piece. Indeed, a pair of ordinary spectacles is nothing less than a pair of small telescopes, for aiding impaired vision.

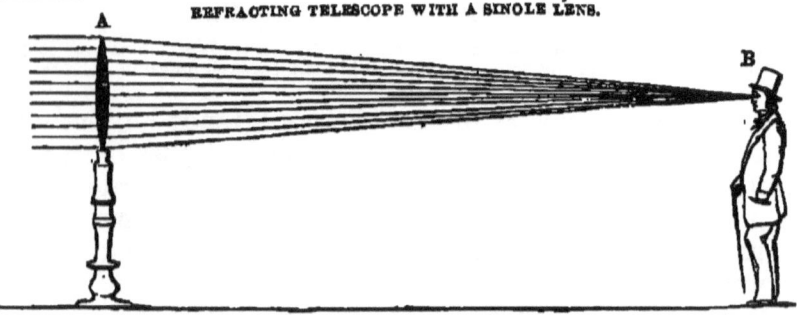

REFRACTING TELESCOPE WITH A SINGLE LENS.

Here the parallel rays are seen to pass through the lens at A, and to be so converged to a point as to enter the eye of the beholder at B. His eye is thus virtually enlarged to the size of the lens at A. But it would be very difficult to direct such a telescope toward celestial objects, or to get the eye in the focus after it was thus directed.

516. The *Galilean telescope* consists of two glasses—a *double-convex* next the object, and a double-concave near the eye. The former converges the light till it can be received by a small double-concave, by which the convergency is corrected (502), and the rays rendered parallel again, though in so small a beam as to be capable of entering the eye.

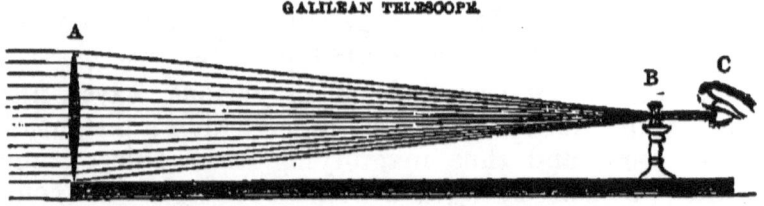

GALILEAN TELESCOPE.

Here the light is converged by the lens A, till it can be received by the double-concave lens B, by which the rays are made to become a small parallel beam, that can enter the eye at C. This was the form of the telescope constructed by *Jansen*, and improved by Galileo; on which account it is called the *Galilean telescope*. In the cut, the two lenses are represented as fastened to a *board*, as first exhibited by Jansen.

517. The common astronomical telescope consists of two glasses—viz., a large double-convex lens next the

---

515. Simplest form of refracting telescope? (Diagram?)
516. *Galilean* telescope? (Diagram and explanation?) Why called *Galilean?*)
517. How common astronomical telescopes made? Why in tube?

object, called the *object-glass;* and a small double-convex lens or microscope next the eye, called the *eye-piece.* For the greater convenience in using, they are both placed in a tube of wood or metal, and mounted in various ways, according to their size, and the purposes to which they are devoted.

LENSES PLACED IN A TUBE.

A is the *object-glass,* B the *eye-piece,* and C the place where the tube in which the eye-piece is set, slides in and out of the large tube, to adjust the eye-piece to the focal distance. By placing the lenses in a tube, the eye is easily placed in the focus, and the object-glass directed toward any desired object.

518. The object-glass of a telescope is usually protected, when not in use, by a brass cap that shuts over the end of the instrument; and the eye-pieces, of which there are several, of different magnifying powers, are so fixed as to screw into a small movable tube in the lower end of the instrument, so as to adjust them respectively to the focus, and to the eyes of different observers. Such telescopes usually represent objects in an inverted position.

REFRACTING TELESCOPE MOUNTED ON A STAND.

The adjoining cut represents the simplest form of a mounted refractor. The object-glass is at A, where the brass cap may be seen covering it. B is the small tube into which the eye-piece is screwed, and which is moved in and out by the small screw C. Two eye-pieces may be seen at D—one short one, for astronomical observations; and a long one, for land objects. For viewing the sun, it is necessary to add a screen, made of colored glass. At E, a bolt goes into a socket in the top of the stand, in which it turns, allowing the telescope to sweep

---

518. How object-glass protected? What said of eye-pieces? (Cut and explanation?)

around the horizon; while the joint, connecting the saddle in which the telescope rests with the top of the bolt, allows it to he directed to any point between the horizon and the zenith. But such stands answer only for comparatively small instruments.

519. **Refracting telescopes** are mounted in various ways. So important is it that they should not shake or vibrate, that, in most observatories, the stand rests upon heavy mason-work in no way connected with the building, so that neither the wind nor the tread of the observer can shake it. They are then furnished with a double axis, which allows of motion up and down, or east and west; and two graduated circles show the precise amount of declination and right ascension. They are then furnished with clockwork, by which the telescope is made to move westward just as fast as the earth turns eastward; so that the celestial object being once found, by setting the instrument for its right ascension and declination, or by the aid of the *Finder*—a small telescope attached to the lower end of the large one—it may be kept in view by the clockwork for any desirable length of time. A telescope thus furnished with right ascension and declination circles is called an *Equatorial*, or is said to be *equatorially mounted*, because it sweeps east and west in the heavens parallel to the equator.

520. The object-glasses of telescopes are not always made of a single piece of glass. They may be made of two concavo-convex glasses, like two watch crystals, with their concave sides toward each other, or with a thin double concave glass between them. They are thus double, or triple; but when thus constructed, the whole is called a *lens*, as if composed of a single piece. Lenses have also been formed by putting two concavo-convex glasses together, and filling the space between them with some transparent fluid. These are called *Barlow* lenses, from Prof. Barlow, their inventor.

521. As a prism analyzes the light, and exhibits different colors, so a double-convex lens may analyze the

---

519. How refractors mounted, and why? When equatorial, and why?
520. How object-glasses made? What a *lens*? A *Barlow* lens?
521. What is an *Achromatic* telescope? (Derivation of term?)

light that falls near its circumference, and thus represent the outside of the heavenly bodies as colored. But this defect is remedied by using disks made of different kinds of glass, so as to correct one refraction by another. Refracting telescopes thus corrected are called *Achromatic* telescopes.

Achromatic is from the Greek *a chroma*, which signifies destitute of color. Most refracting telescopes are now so constructed as to be achromatic.

522. It is but recently that any good refracting telescopes have been made in this country. The best have formerly been made in Germany and France; but they are now manufactured with success, and to considerable extent, by Mr. Henry Fitz, Jun., New York city.

It is now (1858) about seven years since Mr. Fitz commenced the manufacture of refracting telescopes, and thus far he has been very successful. At each fair of the American Institute, for seven years, he has received the highest premium—a gold medal. The glass used by him is obtained from Paris, because none suitable for large telescopes has yet been made in America. His telescopes are perfectly achromatic, and are sold much cheaper than imported ones of the same size and value.

Mr. Fitz has recently made a very valuable improvement in the mounting of telescopes—one which is not only much superior to the old method, but which costs only about one-half as much. This improvement consists in using a single piece of cast-iron in the place of several pieces of brass work. It is very simple, secures great steadiness to the instrument, and is easily adjusted.

The writer is fully satisfied of the value of this improvement, and would recommend it, as well as Mr. Fitz's instruments, to all institutions and amateur astronomers about to purchase either. Besides patronizing a worthy American optician, they will get as good a telescope and much better mounting than by sending abroad, and at far less expense. The following is a list of telescopes manufactured by Mr. Fitz, with the prices attached.

PRICES OF FITZ'S TELESCOPES, EQUATORIALLY MOUNTED, ETC.

Focal length.	Object-glass.	New style mounting.	Old style mounting.	Difference of cost.
9 ft.	8¼ inches.	$1,400	$2,300	$900
11 "	8¼ "		2,220	
10 ft. 8 inch.	8 "	1,500	2,000	850
8 "	6¾ "	700	1,200	500
7 "	5 "	400	750	850
7 "	4½ "	300	500	200
5 "	4 "	225	400	175

He will furnish a very good telescope of three inches aperture for $120, equatorially mounted, with eye-pieces, &c. The size priced at $225 is equal to that at Yale College. A good revolving dome for an observatory building can be built for $100.

This note is inserted exclusively for the benefit of institutions using the work, and without any request or remuneration from Mr. Fitz. Orders or letters of inquiry may be addressed to Henry Fitz, Jun., 237 Third-street, New York.

---

522. Where telescopes formerly made? Where and by whom now, in this country?

RUTHERFORD'S EQUATORIAL REFRACTOR.

523. The above cut represents an equatorial telescope manufactured by Mr. Henry Fitz, of New York—the one used by the author in making most of his observations. Its object-glass is six inches in diameter, and its focal length eight feet.

A is the declination circle, and B the circle of right ascension. The two sticks hanging from these circles are used to move the instrument in right ascension or declination, while the observer is at the eye-piece.

The *Finder* is seen attached to the lower end of the large instrument. It takes in a larger field of view in the heavens than the latter, and enables the observer to look up objects with facility, and bring them into the field of the larger instrument.

This instrument has no clockwork attached. It rests upon a pillar of heavy masonwork, the top of which may be seen in the cut; and in the hands of its present owner, Lewis M. Rutherford, Esq., has already rendered very efficient service.

5:". : r. Rutherford's telescope? By whom made?

DIFFERENT KINDS OF TELESCOPES. 227

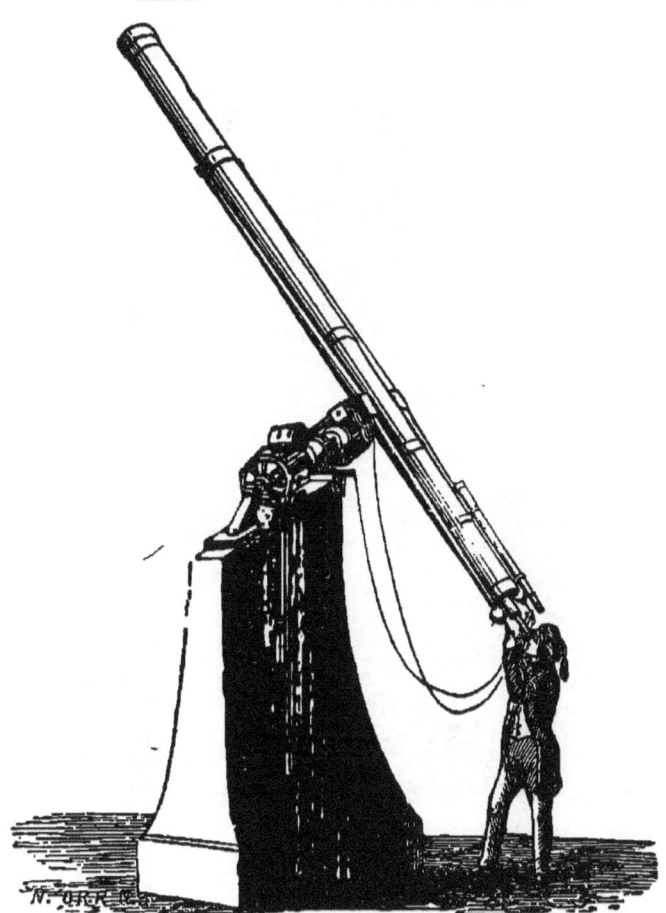

GREAT REFRACTING TELESCOPE AT CINCINNATI, OHIO.

524. The above cut represents one of the most important telescopes in the United States. It is located in the observatory on Mount Adams, near Cincinnati, Ohio, and has been for several years under the direction of *Prof. O. M. Mitchel*, by whose instrumentality it was purchased and mounted.

_{The object-glass is about 12 inches in diameter, with a focal distance of 17 feet. It was purchased in Munich, Germany, in 1844, at an expense of nearly *ten thousand dollars*. There is but one larger than this in the United States, and but two larger in the world.}

---

524. Cincinnati refractor—where located ? By whom purchased ? (Where ? When ? Cost ? Size and focal distance ? Comparative size ?)

228 ASTRONOMY

THE GREAT CRAIG TELESCOPE, WANDSWORTH COMMON, NEAR LONDON.

525. This is the largest refracting telescope ever constructed. The object-glass is two feet in diameter, with a focal distance of 76 feet. The tube is of heavy sheet iron, and shaped somewhat like a cigar. It is 13 feet in circumference in the largest place, and weighs about three tons.

This telescope is suspended from a brick tower 65 feet high, 15 feet in diameter, and weighing 220 tons. The top of the tower, from which the telescope is suspended, revolves; and by a chain running over pulleys, and a weight and windlass, it is balanced, and raised or lowered. The lower end rests on a small carriage, that runs upon a circular railroad around the tower, at the distance of 52 feet from its center. By these means it is directed to almost any point in the heavens. It is called the "Craig" telescope, in honor of Rev. Mr. Craig, under whose direction, and at whose expense, it was constructed. It is located at Wandsworth Common, near London.

525. Describe the Craig telescope. Object-glass?—focal distance? Tube? (How mounted? Why called "Craig" telescope? Where located?)

526. Besides this monster refractor, there are several other very fine instruments in Europe; as the Dorpat telescope, Sir James South's, the Northumberland refractor, the Oxford telescope, &c. Several colleges and seminaries in the United States have observatories connected with them, and telescopes of greater or less value. The largest is at Cambridge, near Boston.

PUBLIC OBSERVATORIES AND TELESCOPES IN THE UNITED STATES

Observatories.	When procured.	Name of maker.	Focal length.	Aperture of object-glass.	Cost.
			ft. in.	inches.	
Yale College....................	1830	Dollond.	10 —	5	$1,000
Wesleyan University..............	1836	Lerebours.	7 —	6	1,000
Williams College................	{ 1836	Holcomb.	10 —	reflector	
	1852	A. Clark.	9 —	7	
Hudson, Ohio....................	1837	Simms.	5 0	4	
Philadelphia....................	1840	Merz.	8 4	6¾	1,900
West Point.....................	1841	Lerebours.	8 —	6	
Washington	1844	Merz.	15 3	9·6	6,000
Cincinnati......................	"	"	17 —	12	9,437
Cambridge	1846	"	22 6	15	19,842
Dartmouth College................	1848	"	9 —	6·4	
Georgetown "	1849	Simms.	7 6	4·8	1,600
Erskine "	"	Fitz.	7 —	5·6	1,050
Shelby "	1850	Merz.	10 4	7·5	8.500
Columbia (S. C.) College...........	1851	Fitz.	8 4	6¾	1,200
Columbia (Mo.) "	1852	"	5 —	4	225

527. Quite a number of very respectable *private* observatories are also in operation in different parts of the country. The following table includes most of them:

PRIVATE OBSERVATORIES.

Private observatories.	When procured.	Maker.	Focal length.	Object-glass.	Cost.
			ft. in.	inches.	
J. Jackson, near Philadelphia...........	1846	—	8 4	6 3-10	$1,833
Mr. Longstreet, Philadelphia...........	"	Fitz.	7 —	5	900
S. G. Gummere, Burlington, N. J.......	1847	"	5 —	4	425
R. Vanarsdale, Newark, N. J...........	{ 1850	"	7 —	5	750
	1851	"	9 4	6¼	1,000
W. S. Van Duzee, Buffalo, N. Y........	"	"	11 —	8½	2,220
W. S. Dickie, Elkton, Ky..............	"	"	6 —	4½	300
D. Mosman, Bangor, Me...............	"	"	5 —	4	225
J. Campbell, New York	1852	"	10 —	8	1,150

526. What other refractors in Europe besides the Craig? Public observatories in this country? Largest telescope? Table.
527. Private observatories—names? Telescopes—by whom mostly made?

528. A *Comet Seeker* is a refracting telescope with a large aperture and short focal distance. As comets cannot be found by their right ascension and declination, but often have to be searched up, by sweeping around the heavens with a telescope, before they become visible to the naked eye, it is important to have telescopes that will cover considerable space—that is, of wide aperture and short focal distance. Such a telescope was made by Mr. Fitz for Miss Mitchel, of Newport, R. I.

A COMET SEEKER.

REFLECTING TELESCOPES.

529. The *Reflecting Telescope* is one in which the light is converged to a focus by means of a concave metallic reflector or speculum. Like the *Refractors*, they may be constructed with very little mounting; though, for convenience in use, it is necessary to place the reflector in a tube.

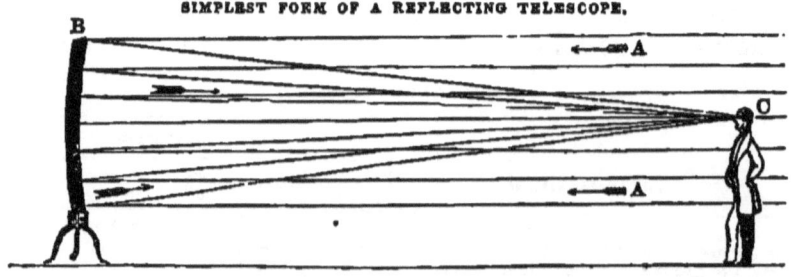

SIMPLEST FORM OF A REFLECTING TELESCOPE.

In this cut, the light A is seen passing from the object on the right, and falling upon the concave surface of the reflector at B, from which it is reflected back to a focus, and enters the eye of the observer at C. This telescope has no eye-piece.

530. The focal distance of a concave reflector is equal to half the radius of the sphere formed by the concave

---

528. What is a *comet seeker?* Why necessary?
529. Describe a reflecting telescope. Simplest form?
530. Focal distance? (Diagram.)

surface produced. Hence to grind a reflector for a focus of 20 feet, it will be necessary to have the curve that of a circle whose radius is 40 feet.

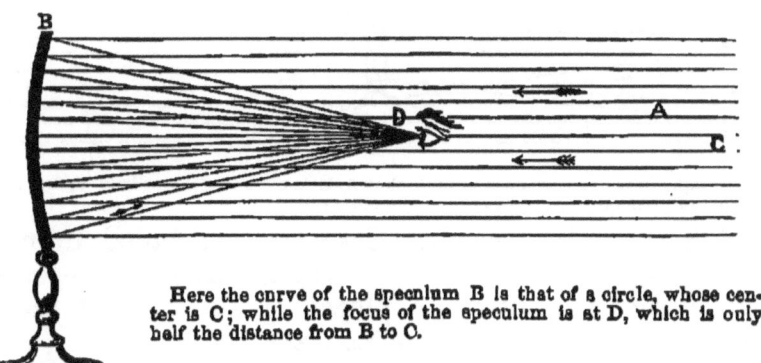

FOCAL DISTANCE OF A CONCAVE REFLECTOR.

Here the curve of the speculum B is that of a circle, whose center is C; while the focus of the speculum is at D, which is only half the distance from B to C.

531. Reflecting telescopes are of several kinds—viz., the *Gregorian*, the *Newtonian*, the *Cassegranian*, the *Herschelian*, &c. The *Gregorian Reflector* has an aperture in the center of the speculum, and a small concave mirror in the focus of the speculum, which reflects the light back through the aperture to the eye-piece. In this way the observer is enabled to face the object, and to direct the telescope toward it, as if it were a refractor.

GREGORIAN REFLECTOR.

Here the light A falls upon the speculum at B, and is reflected back to the small mirror C, by which it is thrown out, through the aperture in the speculum, to the eye of the observer at D. The object is supposed to be off on the right, in the direction toward which the instrument is pointed. It is called a *Gregorian* telescope, after Mr. James Gregory, who first suggested the construction of reflecting telescopes.

532. The *Newtonian Reflector* is so called after Sir Isaac Newton, its inventor. Instead of a concave mirror in the focus of the speculum, he placed a plane mir-

---

531. How many kinds of reflectors? Describe the Gregorian. (Diagram. Why called Gregorian?)
532. Newtonian reflectors? (Diagram and explanation.)

ror there, inclined so as to reflect the light to the *side* of the tube, when it was received by the observer.

NEWTONIAN REFLECTOR.

The light from the speculum is here shown falling upon the inclined mirror in the center, and reflected out to the eye of the observer.

533. The *Cassegranian Reflector* is so called from M. Cassegrain, its inventor. It resembles the Gregorian, except that the speculum placed in the focus of the reflector is *convex*, instead of concave.

534. The *Herschelian Reflector* differs from all others, in having no small reflector whatever; the light being reflected back to a focus at the top of the telescope, and near the edge of the tube, where the eye-piece is placed, and where the observer sits looking into the mirror with his back to the object.

HERSCHELIAN TELESCOPE.

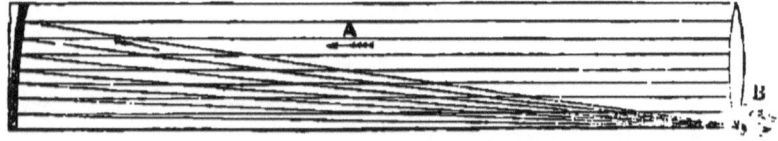

Here the concave speculum is seen to be inclined a little to the lower side of the tube, so that the parallel rays A are reflected back to the observer at B, at the side of the instrument, where the eye-piece is placed.

535. The first telescope constructed upon this plan was that by Sir William Herschel, in 1782. This was called his 20 feet reflector, and was the instrument with which he made many of his observations upon the double stars. In 1789, he completed his *forty feet* reflector, until recently the largest telescope ever constructed.

---

533. Cassegranian? Difference?
534. Herschelian? Where eye-piece? How observer sit?
535. First Herschelian telescope? What called? Next?

SIR WILLIAM HERSCHEL'S FORTY FEET REFLECTOR.

536. The speculum of this instrument is 4 feet in diameter, $3\frac{1}{2}$ inches thick, and weighed, before being ground, 2,118 pounds. The tube is made of sheet iron riveted together, and painted within and without.

<small>The length of the tube is 39 feet 4 inches, and its weight 8,260 pounds. It is elevated or lowered by tackles, attached to strong frame-work; and the observer, who sits in a chair at the upper end of the tube, and looks down into the reflector at the bottom, is raised and lowered with the instrument. Three persons are necessary to use this telescope—one to observe, another to work the tube, and a third to note down the observations. A speaking tube runs from the observer to the house where the assistants are at work. By this telescope, the sixth and seventh satellites of Saturn were discovered; and it was the chief instrument used by its distinguished owner, in making the observations and discoveries which have immortalized his name, and which have so abundantly enriched and advanced the science of astronomy.</small>

---

536. Herschel's forty feet reflector? Size of speculum? Weight? Tube? Length and weight? How mounted? Observer where? Usefulness?

LORD ROSSE'S GREAT REFLECTING TELESCOPE.

537. This is the largest reflecting telescope ever constructed. The speculum, composed of copper and tin, weighed three tons as it came from the mould, and lost about ⅛th of an inch in grinding. It is 5½ inches thick, and 6 feet in diameter. It was cast on the 13th of April, 1842, and was cooled gradually in an oven for 16 weeks, to prevent its cracking, by a sudden or unequal reduction of the temperature. This speculum has a reflecting surface of 4,071 square inches. The tube is made of deal wood, one inch thick, and hooped with iron. Its diameter is seven feet, and its length 56.

The entire weight of this telescope is twelve tons. It is mounted between two north and south walls, 24 feet apart, 72 feet long, and 48 feet high. The lower end rests upon a universal hinge. It can be lowered to the horizon, and raised to the zenith, and lowered northward till it takes in the Pole star. Its motion from east to west is limited to 15 degrees. This magnificent instrument is situated at Burr Castle, Ireland. It was constructed by the Earl of Rosse, at an expense of $60,000.

537. Lord Rosse's telescope? Weight of speculum? Diameter? Thickness? Cooling? Tube? Entire weight? How mounted? What motion? Where located? Cost?

A TRANSIT INSTRUMENT.

538. A *Transit Instrument* is a telescope used for observing the transit of celestial objects across the meridian, for the purpose of determining differences of right ascension, or obtaining correct time. They are usually from six to ten feet long, and are mounted upon a horizontal axis, between two abutments of mason work; so that the instrument, when horizontal, will point exactly to the south. It will then take objects in the heavens, when they are exactly on the meridian.

Let A D in the cut represent the telescope, and E and W the east and west abutments, between which it is placed. On the left is seen, attached to the mason work, a graduated circle; and on the eastern end of the axis of the telescope is seen an arm, n, extending to the circle, as an index. Now, suppose the index n to be at o, in the upper part of the circle, when the telescope is horizontal; then if the meridian altitude of the object to be taken is 10°, the index must be moved 10° from o, as the degrees on the circle and the altitude of the object will correspond.

---

538. What is a transit instrument? Size? How mounted? (Describe parts as shown in the cut. How set the instrument for the meridian altitude of a star?)

236 ASTRONOMY.

539. An *Astronomical Clock* is a clock adapted to keep exact sidereal time (136).

Taking the vernal equinox in the heavens as the zero point, and reckoning 24 hours eastward to the same point again, the time—reckoning 15° to an hour—when an object crosses the meridian, will always represent the right ascension of the object. Hence right ascension is usually given in hours, minutes, and seconds; or in time by the astronomical clock, set by the vernal equinox.

THE MURAL CIRCLE.

540. A *Mural Circle* is a large graduated circle, with a telescope crossing its center, used for the measurement of the altitudes and zenith distances of the heavenly bodies, at the instant of their crossing the meridian. They are usually fixed upon a horizontal axis, that turns in a socket firmly fixed in a north and south wall. The degrees, minutes, and seconds on the circle are read by means of microscopes, and indicate the altitude of the object.

In the cut, A is a reading microscope, and B C D E the wall to which the circle is attached. The telescope would denote an altitude of about 40°, which would leave 50° as the zenith distance.

539. An astronomical clock? How set? How indicate right ascension of objects?
540. Describe a mural circle? Its uses? How mounted? (How ascertain altitude and zenith distance by it?)

541. Parallax is the difference between the altitude of any celestial object seen from the earth's surface, and the altitude of the same object seen at the same time from the earth's center; or it is the angle under which the semi-diameter of the earth would appear, as seen from the object.

The *true place* of a celestial body is that point of the heavens in which it would be seen by an eye placed at the center of the earth. The *apparent place* is that point of the heavens where the body is seen from the surface of the earth. The parallax of a heavenly body is greatest when in the horizon, and is thence called the *horizontal parallax*. Parallax decreases as the body ascends toward the zenith, at which place it is nothing.

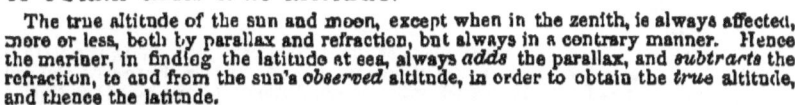

PARALLAX OF THE PLANETS.

The adjoining cut will afford a sufficient illustration. When the observer, standing upon the earth at A, views the object at B, it appears to be at C, when, at the same time, if viewed from the center of the earth, it would appear to be at D. The parallax is the angle B C D or A B E, which is the difference between the altitude of the object B, when seen from the earth's surface, and when seen from her center. It is also the angle under which the semi-diameter of the earth, A E, is seen from the object B.

As the object advances from the horizon to the zenith, the parallax is seen gradually to diminish, till at F it has no parallax, or its apparent and true place are the same.

This diagram will also show why objects *nearest* the earth have the *greatest* parallax, and those most distant the least; why the moon, the nearest of all the heavenly bodies, has the greatest parallax; while the fixed stars, from their immense distance, have no appreciable horizontal parallax—the semi-diameter of the earth, at such a distance, being no more than a point.

542. As the effect of parallax on a heavenly body is to depress it *below its true place*, it must necessarily affect its right ascension and declination, its latitude and longitude. On this account, the parallax of the sun and moon must be *added* to their *apparent* altitude, in order to obtain their *true* altitude.

The true altitude of the sun and moon, except when in the zenith, is always affected, more or less, both by parallax and refraction, but always in a contrary manner. Hence the mariner, in finding the latitude at sea, always *adds* the parallax, and *subtracts* the refraction, to and from the sun's *observed* altitude, in order to obtain the *true* altitude, and thence the latitude.

541. Parallax? True place of a celestial body? Apparent? When parallax greatest? Least? Called what, and why? (Diagram? What objects greatest parallax?)

542. Effect of parallax? How obtain true altitude? (How differ from refraction? How then obtain true altitude?)

543. The principles of parallax are of great importance to astronomy, as they enable us to determine the distances of the heavenly bodies from the earth, the magnitudes of the planets, and the dimensions of their orbits.

The sun's horizontal parallax being accurately known, the earth's distance from the sun becomes known; and the earth's distance from the sun being known, that of all the planets may be known also, because we know the exact periods of their sidereal revolutions, and, according to the third law of Kepler, the squares of the times of their revolutions are proportional to the cubes of their mean distances. Hence, the first great desideratum in astronomy, where measure and magnitude are concerned, is the determination of the true parallax.

<small>At a council of astronomers assembled in London some years since, from the most learned nations in Europe, the sun's mean horizontal parallax was settled, as the result of their united observations, at 0° 0' 8".5776. Now the value of radius, expressed likewise in seconds, is 206264".8; and this divided by 8".5776, gives 24047 for the distance of the sun from the earth, in semi-diameters of the latter. If we take the *equatorial* semi-diameter of the earth as sanctioned by the same tribunal, at (7924÷2=) 3962 miles, we shall have 24047×3962=95,273,869 miles for the sun's true distance.</small>

544. The change in the apparent position of the fixed stars, caused by the change of the earth's place in her revolution around the sun, is called their *annual parallax*. So immense is their distance, that the semi-annual variation of 190,000,000 of miles in the earth's distance, from all those stars that lie in the plane of her orbit, makes no perceptible difference in their apparent magnitude or brightness.

<small>The following cut will illustrate our meaning:</small>

<small>Let A represent a fixed star in the plane of the earth's orbit, B. At C, the earth is 190,000,000 of miles nearer the star than it will be at D six months afterward; and yet this semi-annual variation of 190,000,000 miles in the distance of the star is so small a fraction of the whole distance to it, as neither to increase or diminish its apparent brightness.</small>

<small>543. Use of parallax? How employed? (Note?)
544. What meant by earth's *annual parallax?* Effect of variation of earth's distance on the fixed stars? (Diagram.)</small>

545. It is only those stars that are situated near the axis of the earth's orbit whose parallax can be measured at all, on account of its almost imperceptible quantity. So distant are they, that the variation of 190,000,000 miles in the earth's place causes an apparent change of less than 1' in the nearest and most favorably situated fixed star.

PARALLAX OF THE STARS.

<small>Let A represent the earth on the 1st of January, and B a star observed at that time. Of course, its apparent place in the more distant heavens will be at C. But in six months the earth will be at D, and the star B will appear to be at E. The angle A B D or C B E will constitute the parallatic angle. In the cut, this angle amounts to about 48°, whereas the real parallax of the stars is less than $\frac{1}{80}$th of one degree, or $\frac{1}{30000}$th part this amount. Lines approaching each other thus slowly would appear parallel; and the earth's orbit, if filled with a globe of fire, and viewed from the fixed stars, would appear but a point of light 1' in diameter!</small>

MISCELLANIA.

546. The *Atmosphere* is an elastic gas, which surrounds the earth on every side. It is supposed to be from 40 to 60 miles in hight, growing more rare as we ascend, and is kept around the earth by attraction.

547. *Wind* is air put in motion by heat, causing bodies of air to rise from the earth's surface, and other air to rush in to supply its place. The velocity of the wind ranges from 5 to 100 miles an hour.

548. *Clouds* are collections of vapor suspended in the air. They range from two miles to half a mile in hight, according to their density and weight. They serve to screen us from the oppressive heat of the sun, and to convey water from the rivers and oceans, and pour it down in showers upon the earth.

549. *Rain* is water condensed, or collected into drops by attraction, and falling from the clouds. *Hail* is drops

---

<small>545. What stars have perceptible parallax? Amount? (Diagram, and explain.)
546. What is the atmosphere? Extent? How kept around the earth?
547. Wind? How put in motion? Velocity?
548. Clouds? Uses?
549. Rain? Hail? Snow?</small>

of rain frozen on its way to the earth; and *Snow* is particles of clouds frozen before being condensed into drops.

550. *Lightning* is electricity passing from one cloud to another, or between the clouds and the earth; and *Thunder* is the sudden shock given to the atmosphere by the passage of the electricity through it.

551. The *Aurora Borealis*, or Northern Light, is a reddish unsteady light sometimes seen in the north. It is supposed to be caused by electricity passing through the upper regions of the atmosphere, about the North Pole.

552. "*Shooting Stars*" are meteors that shoot downward toward the earth, like stars falling from their spheres. They are usually seen one at a time, and only in the night, but sometimes fall in showers, and no doubt fall in the day time, though invisible.

From 2 o'clock in the morning, November 13, 1833, till daylight, the whole heavens were filled with these fiery particles and streaks of light darting downward from the sky. These meteors, no doubt, come from regions beyond the limits of the atmosphere, and are ignited by their rapid passage through it. Their origin and nature are as yet matters of inquiry and speculation.

553. *Aerolites*, or Meteoric Stones, are masses of stone or iron that have fallen from the sky at various periods, and on almost every part of the globe. They are often found after the explosion of large meteors, sometimes while they are yet hot.

A large meteor exploded over Cabarras county, North Carolina, a few years since, several pieces of which were picked up the next day. One piece, weighing 19 lbs., had struck a large pine tree lying on the ground, and had gone through it, and into the earth, to the depth of three feet. In some cases, large masses of iron have fallen. In December, 1795, a stone weighing 51 lbs. fell in Yorkshire, England. The writer has a piece of an aerolite that weighed 90 lbs., that fell in New Jersey. A large mass of meteoric iron may be seen in the museum of Yale College.

550. Lightning and thunder?
551. Aurora Borealis?
552. "Shooting stars?" How seen? (What shower mentioned? Distance from which they come?)
553. Aerolites? (What instances of their fall cited?)

www.ingramcontent.com/pod-product-compliance
Lightning Source LLC
Chambersburg PA
CBHW031745230426
43669CB00007B/490